Advances in Geographic Information Science

More information about this series at http://www.springer.com/series/7712

Shivanand Balram • James Boxall
Editors

GIScience Teaching and Learning Perspectives

Springer

Editors
Shivanand Balram
Department of Geography
Simon Fraser University
Burnaby, BC, Canada

James Boxall
GIS Centre, Killam Library
Dalhousie University
Halifax, NS, Canada

ISSN 1867-2434 ISSN 1867-2442 (electronic)
Advances in Geographic Information Science
ISBN 978-3-030-06057-2 ISBN 978-3-030-06058-9 (eBook)
https://doi.org/10.1007/978-3-030-06058-9

Library of Congress Control Number: 2019930432

This Springer imprint is published by the registered company Springer Nature Switzerland AG
The registered company address is: Gewerbestrasse 11, 6330 Cham, Switzerland

Foreword

GIS is undergoing a major paradigm shift. Not only are the tools and data migrating to the web becoming accessible on an ever-growing array of mobile devices and in many forms of multimedia, but the ability to share these tools and data sets is bringing fundamental shifts to how geospatial information is used and valued throughout society. This Web GIS paradigm also brings to focus fundamental questions to those involved in GIS education: What content should we teach to prepare students for twenty-first-century society? What skills should we foster to enable graduates to find secure work and to make a meaningful contribution to society through GIS? How should we teach this content and these skills? What tools and methods should we use to teach these content and skills? What are some meaningful case studies, and what trends should I be examining as an instructor to stay current and to keep my students moving forward?

At such a time, the book *GIScience Teaching and Learning Perspectives* is a valuable and practical contribution to GIS and to GIS education. While focused on the Canadian higher education context, the themes of the book are global in nature and include important questions such as: What constitutes innovative pedagogy? How can research engage real-world issues in the community? What changes, if any, need to be made so that GIScience can contribute to the wider academy?

It is fitting that this book focuses on GIScience in higher education in Canada. It was in Canada during the 1960s where Dr. Roger Tomlinson created the first practical GIS at the Canada Land Inventory. It was in Canada where Dr. Michael Goodchild during the 1980s laid the groundwork for transforming GIS from a "system" or a set of tools into a "science" – GIScience – encompassing theoretical foundations. This book does not treat education as less important than other applications of GIScience or its theoretical underpinnings. On the contrary, as reflected in the book's title, and in the first chapter where Shivanand Balram sets the tone, in *Teaching and Learning Pedagogies in Higher Education Geographic Information Science*, the teaching of GIScience is integrally tied to understanding GIScience and its advancement.

This book brings together some of the most innovative and knowledgeable people in Canada to provide the current state of research and development in higher education. They hail from such universities with long-standing GIS programs such as Simon Fraser, Dalhousie, Alberta, Lethbridge, Mount Royal, and elsewhere. Yet the authors are not content to describe the "current state," but each looks forward to what is *possible* in the future of GIS in education. Any reader will be left in amazement that so much progress has been made in a field that is only 50 years old but also is left with the keen impression that many stones are barely touched or left unturned. Thus, the book encourages the reader to imagine the possibilities of what *could be* and then consider actions that could be taken to help achieve societal goals of GIS in education. These goals include empowering students with confidence and skills to create smart cities and sustainable futures that will ensure healthy ecosystems and cities. To accomplish this, the authors "tell it like it is." They don't mince words: if something is amiss with the current situation of GIS and GIScience, they clearly state what it is and why it is that way. James Boxall, for example, in his chapter *Epilogue: A Future of Convergence and Crises* (Chap. 10), discusses the real tensions that sometimes exist between the promotion of geography and geomatics in higher education. Yet his hopeful tone shines through that current efforts to spread spatial thinking and geotechnologies throughout the educational system are "not too late – just too little." Hall-Beyer, in *Using an Online Format to Teach Graduate-Level Remote Sensing Basics* (Chap. 6), is also hopeful in her reminder that "GIS above all fields is able to provide unique insights."

This book is practical – providing sound advice for deans, provosts, and especially educators to implement in their instruction. Lynn Moorman's focus in *The Evolution and Definition of Geospatial Literacy* (Chap. 2) is on teaching with GIS with a goal of moving students from basic skills of reading and writing to functional skills that are measurable technical skills. These skills go beyond education – indeed, Dr. Moorman makes the case that these are *life skills*. She takes it even further: these skills are *multiliteracies* – they draw from many types of hard and soft skills and contribute to many disciplines, from public safety to health, to energy, to hydrology, to urban planning, and more. Corbett and Legault, in *Neogeography: Rethinking Participatory Mapping and Place-Based Learning in the Age of the Geoweb* (Chap. 8), study how place-based education can be effectively taught using Web GIS tools and approaches. Dragićević and Anderson in their well-designed study on *Enabling Scientific Research Skills in Undergraduate Students During a Spatial Modeling Course* (Chap. 3) report evidence on the importance of developing scientific research skills in GIS students. Huynh and Hall, in *Navigating Employment Prospects for New Graduates in the Geospatial Sciences* (Chap. 9), provide a set of skills assessments that instructors and students could use as is or modify to suit their needs. Students are provided work sheets to assess and keep track of their contacts. The authors do this because they believe that "growing your network" is key to success in the field.

The book is filled with interesting and useful case studies. Craig Coburn, in *The AMETHYST Program: The NSERC CREATE Experience* (Chap. 4), discusses the Advanced Methods, Education and Training in Hyperspectral Science and Technology program at the University of Lethbridge. This program achieved success in student professional development and collaborative experiences despite operating in a period of diminished funding in higher education science. Han, in *Web GIS in Development: From Research and Teaching Perspectives* (Chap. 7), analyzes the development of Web GIS in Canada and its profound impact on teaching and learning, including its challenges and its benefits. His case study, "Mapping the Media in the Americas," investigates the relationship between media and democracy in 11 Latin American countries. Huynh and Hall, in *Navigating Employment Prospects for New Graduates in the Geospatial Sciences* (Chap. 9), weave an excellent synthesis of a wide variety of studies in Canada and the USA from leaders in higher education, industry, and nonprofit organizations that have examined workforce skills, graduation statistics, and trends and programs in higher education.

Geoliteracy is a theme that runs through many of the chapters in the book. Authors such as Moorman (Chap. 2) and Boxall (Chap. 10) draw from their long, rich experience as educators to make the case that Geoliteracy is more than content knowledge or skills. Rather, Geoliteracy is also intricately tied to the geographic perspective and is shaped by individual researchers as well as by the pedagogical and administrative approaches that the instructor uses in the higher education teaching and learning context.

Workforce development is another theme that runs through the book. Coburn (Chap. 4) discusses how the University of Lethbridge program established pathways for students to gain experience in international organizations as interns and as employees. Dragićević and Anderson (Chap. 3) examine the role that scientific research skills play in students pursuing graduate studies. This study is important because if the field of GIScience is to develop, we need to train the next generation of GIScience researchers. Huynh and Hall (Chap. 9) analyze this theme on a broader scale, examining the global geospatial industry, employment prospects for Canadian graduates, and skills necessary to thrive within the industry. Their work addresses technical skills and the "soft skills" of presenting, communicating, teamwork, arguing a point, and others in a chapter nicely accented with anecdotes of personal job search strategies. They examine key initiatives that sought to identify competencies, including the Geospatial Technology Competency Model and workplace surveys from the Urban and Regional Information Systems Association (URISA).

This book is also *innovative* – breaking new ground in GIScience research. Lynn Moorman (Chap. 2) raises the questions of whether and how dispositions – natural or acquired combinations of qualities that a person demonstrates – affect a person's ability to think spatially and work with geospatial technologies. In *Using an Online Format to Teach Graduate-Level Remote Sensing Basics* (Chap. 6), Mryka Hall-Beyer advocates that undergraduate GIS training should establish outcomes that guide the student through increasing technical expertise. Her intriguing title "The Map Is Not the Territory" in Chap. 5 is meant to convey that "with only technical emphasis in GIS, there is a temptation to mistake added data quantity and algorithm complexity (GIS's 'map') for the 'territory' of the area to which it is being applied." In other words, adding "more tech" goes beyond increasing technical skills. It enhances the student's ability to innovate and communicate. The book also includes innovative teaching methods, such as Coburn's discussion of blended learning and Corbett and Legault's discussion on how to engage students in civic involvement, where they state "However, we should not downplay the importance of these types of projects to contribute to building community cohesion, as well as the broader university student base, to engage in place-related issues; to raise awareness about pressing land related issues; and ultimately contribute to empowering individuals to take a greater role in seeking solutions to these issues." Dragićević and Anderson (Chap. 3) outline an innovative structuring of various pedagogical tools to facilitate scientific research skills in undergraduate spatial modeling students.

For readers seeking background on how GIS and GIS in education evolved since the 1960s, Hall-Beyer (Chap. 6) provides an excellent foundation and grapples with the continued reexamination of the definition and use of the term "GIS." But the book is also forward-looking in light of the paradigm shift to Web GIS. Corbett and Legault (Chap. 8) take participatory mapping and place-based learning, two fairly long-standing aims of geographic teaching and learning, into the age of the "geoweb" or Web GIS. They anchor their study into new approaches to geography sometimes called "neogeography" where complex techniques of cartography and GIS are put within reach of users and developers. Their chapter includes some intriguing participatory GIS work in studying the proposed Northern Gateway Pipeline in Alberta and the Okanagan Fruit Tree Project that harvests fruit from backyard trees to reallocate produce to those in need in the Kelowna community.

In today's rapidly changing digital educational environment, educators and others need practical advice and best practices. The book deals squarely with online instructional strategies. Many provide guidance for teaching and learning in online environments in higher education, including via online courses, open textbooks, and using open software, with a case study of students involved with an 8-week online course. Hall-Beyer (Chap. 6) presents research based on six offerings of an online course in remote sensing stretching back to 2010. Blended learning and flipped classrooms are also critically analyzed in the book.

The book reminds the reader that teaching with GIS is a practical endeavor but one with big goals. For example, the *Mapping Media* project developed by Ruibo Han, in *Web GIS in Development: From Research and Teaching Perspectives* (Chap. 7), was conducted not only to support the integration of maps in fact-based research but rather to strengthen the technical capacity of Latin American partners and to promote the maps among a wide range of stakeholders as a resource to further explore the realities of how media impacts democracy and vice versa.

Another of the book's themes is how to evaluate success in GIS education. Again, the authors challenge our thinking: "There is little evidence to support the notion that the traditional student/professor mentor model of graduate student training (also a form of experiential learning) is ineffective at providing relevant job skills for our students." Hall-Beyer reminds us that "reality always has more about it than we can capture in our data systems and more than is dreamed of in our analysis and output." Also, Dragićević and Anderson make the case for higher-order thinking through the research process, and they demonstrate the notion in a spatial modeling course. The message for instructors (and students!) is think beyond the map. The book's authors are also critically reflective of their own work as they encourage students to do the same.

The book also provides a welcome context, anchoring GIS education with broader technological and societal forces. In the final chapter, Boxall (Chap. 10) provides an inspiring, breathtaking tour of over a century of these forces, why they matter, and how they can and should impact our current perspective and instruction. This book reminds us in GIS education that while the road may be unclear and rocky at times, it is a noble and worthy endeavor that we are on – increasing spatial literacy for the planet, and ultimately, so that the planet and its people will benefit in positive ways.

Joseph J. Kerski
Esri and University of Denver
Denver, CO, USA

Acknowledgments

Developing a book is a team effort, and we are indebted to many persons for their assistance throughout the journey. First, we thank all the contributing authors for their time, effort, and research ideas presented in the individual chapters. Collectively, these chapters give a wonderful snapshot of the state and future potential of GIScience teaching and learning focusing on the Canadian experience.

All the chapters were peer-reviewed, and the chapter authors have updated their work to address the feedback of the reviewers. We thank all the anonymous reviewers who gave their time to provide feedback for the chapters.

We are grateful to Dr. Joseph J. Kerski (Esri and University of Denver, USA) for agreeing to write the Foreword and to continuously edit it as we kept adding new chapters and/or changing chapter titles in the book.

Shivanand Balram thanks all his GIScience research colleagues, teachers, supervisors, and students for their interactions that today help shaped his own perspective on the teaching and learning of GIScience in higher education.

James Boxall wishes to thank his colleagues and students for their patience in having to accommodate the never-ending questions about their viewpoints. As always, the deepest gratitude to the many mentors with whom he has had the good fortune to listen and gather the best advice anywhere (next step is to follow it).

Finally, the Springer Management Team (Ron Doering, Aaron Schiller, Chandhini Kuppusamy, and Susan Westendorf) was especially accommodating to us, and those accommodations ensured that the book was completed successfully. It was a pleasure to work with them throughout the process.

Shivanand Balram and James Boxall

Contents

Contributors

Taylor Anderson Spatial Analysis and Modeling Research Laboratory, Department of Geography, Simon Fraser University, Burnaby, BC, Canada

Shivanand Balram Department of Geography, Simon Fraser University, Burnaby, BC, Canada

James Boxall Dalhousie University, GISciences Centre, Killam Library, Halifax, NS, Canada

Craig Coburn Department of Geography, University of Lethbridge, Lethbridge, Canada

Jon Corbett Community, Culture and Global Studies, UBC Okanagan, Kelowna, BC, Canada

Suzana Dragićević Spatial Analysis and Modeling Research Laboratory, Department of Geography, Simon Fraser University, Burnaby, BC, Canada

G. Brent Hall ESRI, Toronto, ON, Canada

Mryka Hall-Beyer Department of Geography, University of Calgary, Calgary, AB, Canada

Ruibo Han Department of Geographical Sciences, University of Maryland, College Park, MD, USA

Niem Tu Huynh Independent scholar, Montréal, QC, Canada

Gabrielle Legault Community, Culture and Global Studies, UBC Okanagan, Kelowna, BC, Canada

Lynn Moorman Mount Royal University, Calgary, Canada

Teaching and Learning Pedagogies in Higher Education Geographic Information Science

Shivanand Balram

Abstract Geographic information science (GIScience) has evolved into an almost universally applicable body of knowledge and can be found in the curriculum of a wide range of disciplines. In the teaching and learning of GIScience, the traditional lecture-based pedagogies and the newer experiential learning pedagogies are the primary approaches. In this study, a background review was conducted focusing on the experiential pedagogies used to teach and learn GIScience in higher education contexts. However, given the complex learning needs of the modern university student, it may be that neither lecture-based nor experiential learning alone would be effective, and they need to be combined in innovative ways to enhance student learning. Consequently, this study provides higher education instructors with an overview of multiple pedagogies that can be used in various combinations to motivate and engage students in a flexible learning process.

Keywords Experiential learning · Flexible learning · GIScience · Pedagogy

Introduction

Geographic information science (GIScience) is the concepts, methods, and computer tools (GISystems) related to the use of geographically referenced data for problem-solving and analysis. The teaching and learning of GIScience have now become a central component of curriculum design and planning in higher education contexts. This focus on teaching and learning has been increasingly necessary because of (1) the ubiquitous nature of GIScience due to its universal applicability and relevance, (2) the need for both knowledge-based and problem-solving skills in learners, and (3) the wide range of pedagogies being used in the classroom and the need to determine their fitness for use in various contexts (Baker et al. 2015; Solari and Schee 2015; Unwin et al. 2011). In a general sense, pedagogy is usually

S. Balram (✉)
Department of Geography, Simon Fraser University, Burnaby, BC, Canada
e-mail: sbalram1@sfu.ca

S. Balram, J. Boxall (eds.), *GIScience Teaching and Learning Perspectives*, Advances in Geographic Information Science,
https://doi.org/10.1007/978-3-030-06058-9_1

taken to mean the systematic set of guiding principles and techniques individual instructors use to achieve well-defined teaching and learning outcomes. The teaching and learning outcomes at the course level determine the skills and competencies of students at the program level of which the courses form a part. Hence, pedagogies are the crucial link between courses and programs in higher education. At a general level, pedagogies can be classified into the two general categories: (1) lecture-based and (2) experiential.

Lecture-based pedagogies assume that the instructor is the authoritative repository of information that is transmitted in one direction from the instructor to the learner. Teaching and learning occur through prepackaged materials such as lectern speeches, blackboard and chalk illustrations, assignment tasks, and formative and summative assessments. In this approach, the learner is a passive recipient of information which leads to many disadvantages including (1) overloading of theory at the expense of practical skills, (2) minimal interaction and engagement between learners and between instructor and learners, and (3) limited opportunities to develop critical thinking and problem-solving skills. Improvements to the traditional lecture-based pedagogy has largely taken the form of using information and communication technologies (ICT) to address shortcomings and focus more on the learner (Balram and Dragicevic 2005, 2008). This newer digital learning approach assumes the teacher is a facilitator and has been implemented in forms such as computer-assisted learning, flipped classroom, blended learning, distance education, e-learning, online learning, and quite recently massive open online courses (MOOCs) (Burrows et al. 2013; Clark et al. 2007).

Experiential pedagogies place the instructor in a facilitator mode and focus on well-designed instructional materials to facilitate student learning, engagement, and reflection on the subject matter. The main guiding principles of experiential learning pedagogies include the following: (1) direct engagement with the subject matter, (2) use a problem-solving process to structure the inquiry, (3) address real-world local and global problems, and (4) use multiple problem contexts to facilitate interdisciplinary learning. The experiential learning approach has been implemented in forms such as active learning, problem-based learning, project-based learning, service learning, and place-based learning.

In this study, a background literature review was conducted to characterize the dominant experiential pedagogies being used to teach and learn GIScience in higher education. The results can provide higher education instructors with an overview of multiple pedagogies that can be used in various combinations to effectively motivate and engage GIS students in a flexible learning process.

Experiential Learning Approaches

Active Learning

The goal of active learning (AL) is to facilitate learner participation and interaction in the classroom (Fig. 1). The participation and interaction can occur on the levels of the individual, in pairs, in small groups, and in large groups involving the entire

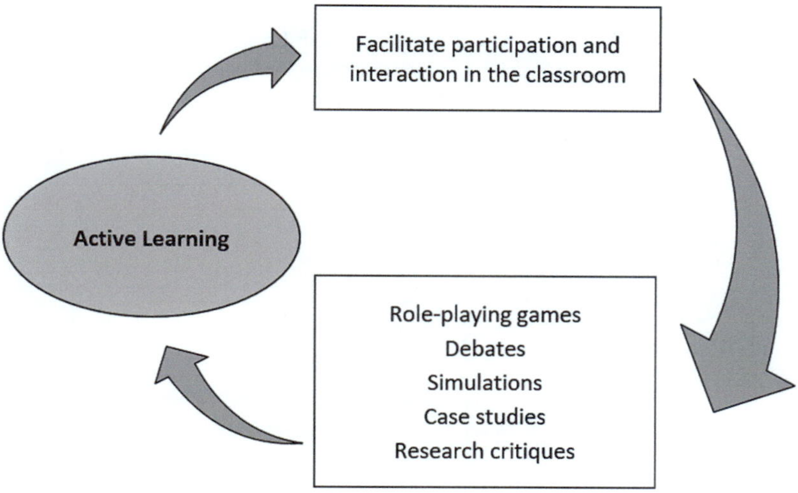

Fig. 1 Main components of the active learning process

set of learners. Another important feature of active learning is that there is some level of reflection or thinking about the activities being done or observed. Furthermore, this reflection or thinking process allows the learners to articulate the content into finding successful solutions to real-world societal problems.

In active learning, the participation and interaction usually take the form of discussion strategies such as role-playing games, debates, simulations, case studies, and research paper critiques. Despite the benefits, there are some drawbacks that need to be considered when implementing active learning strategies. One important drawback is the imposter syndrome where some learners upon hearing the discussion responses of others feel they are not smart enough and don't belong to the discussion or in the classroom. This situation can be alleviated by the instructor taking the necessary steps to create a safe learning environment so that every learner contributes in some way to the active learning strategies.

Problem-Based and Inquiry-Based Learning

The goal of problem-based learning (PBL) is to define and implement an in-depth investigation that leads to feasible solutions to a real-world problem (Fig. 2). In the process, learners actively seek out and apply knowledge from multiple disciplines and articulate those into feasible solutions (Drennon 2005; King 2008; Kinniburgh 2010). The instructor acts as a coach and guides students as they develop critical thinking, problem-solving, and collaboration skills toward their feasible problem solution. Inquiry-based learning (IBL) is similar to PBL except for how the learning problem is defined. In PBL, the instructor outlines the specific problem to be solved, while in IBL learners select and specify their own problems to solve.

In PBL/IBL, the learning is loosely structured around the scientific method. In the first stage, learners do a scoping review, assess relevant research work in the

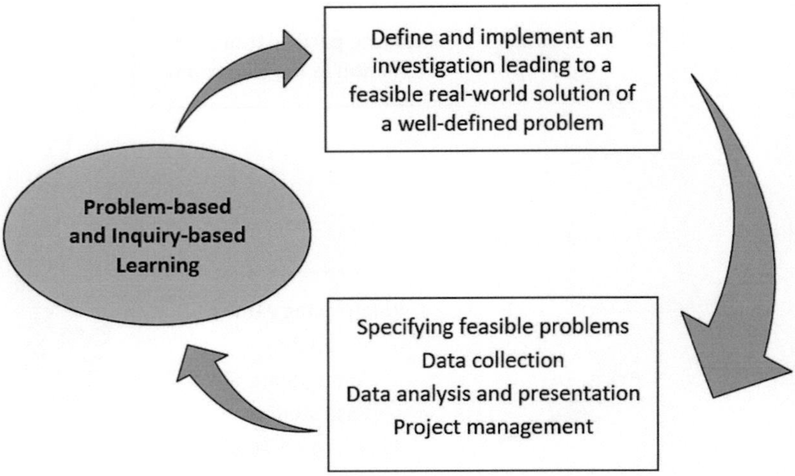

Fig. 2 Main components of the problem-based and inquiry-based learning process

scientific literature, examine the problem to be solved, and define techniques and data needed for a feasible solution. In the second stage, learners will plan and implement their solution strategy. During this stage, the work may be divided up into components so that each learner contributes to the solution. In the final stage, learners integrate their components and shared their results with each other to create the final feasible solution. One important drawback is the problem definition phase. Learners who are not advanced in thinking and reasoning skills may encounter difficulty in this phase, and so one solution is for PBL to be used for new learners in the subject matter and for IBL to be used for relatively advanced learners.

Project-Based Learning

The goal of project-based learning is to provide learners with an opportunity to take ownership of the learning process by allowing them to design and develop projects that would yield meaningful personal learning experiences (Fig. 3). While the process shares many similarities with PBL and IBL, one distinguishing feature is that in project-based learning, students may require multiple attempts to solve a problem to their satisfaction. Consequently, time constraint is usually one of the primary drawbacks. However, it can be alleviated by controlling the instructor input and the learner input in defining the problem and solution stages such that the completion time for solution is managed much more effectively.

The project-based learning is structured broadly by categories such as problem definition, data collection and processing, and output products. These are all tightly fitted within the traditional project management framework with tasks to be done, timelines for completion of tasks, and resources allocated to each of these tasks.

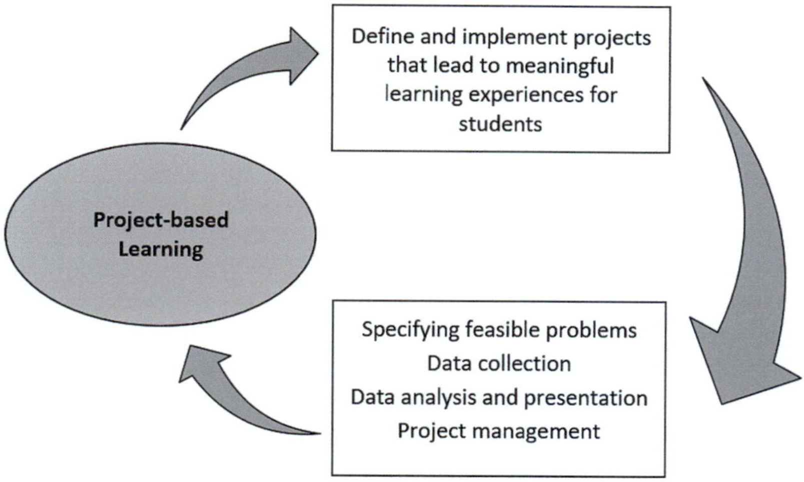

Fig. 3 Main components of the project-based learning process

Usually the entire project-based learning process is assessed based on a project proposal, output products such as reports and maps, and a public presentation to communicate the solutions to the problem.

Service Learning

The goal of service learning is to embed community work into academic learning in ways that are mutually reinforcing (Fig. 4). The process shares many similarities with PBL and IBL and in general occurs by identifying a community need, developing a solution plan based on academic knowledge, implementing the solution in collaboration with the community, reflection on the entire process, and public presentation to communicate results and share learning experiences (Gordon et al. 2016). The study abroad programs of many universities have been a popular means to facilitate global service learning opportunities (Doerr 2013). The instructor may serve as a guide and mentor in cases where learners may be relatively advanced in the subject matter or as a direct participant in the activities to motivate students.

Place-Based Learning

The goal of place-based learning is to anchor the learning within the local context to enhance both the local environment as well as student learning and achievement (Elwood and Wilson 2017; Tate and Jarvis 2017) (Fig. 5). Place-based learning shares many similarities with service learning in the sense that they focus on

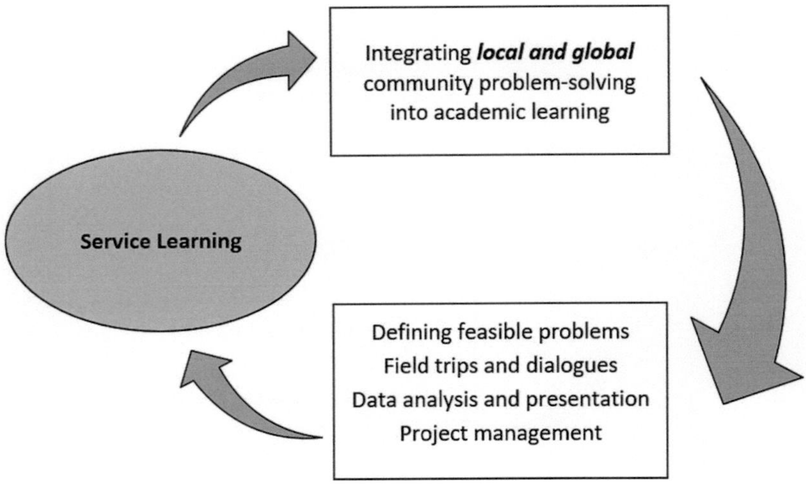

Fig. 4 Main components of the service learning process

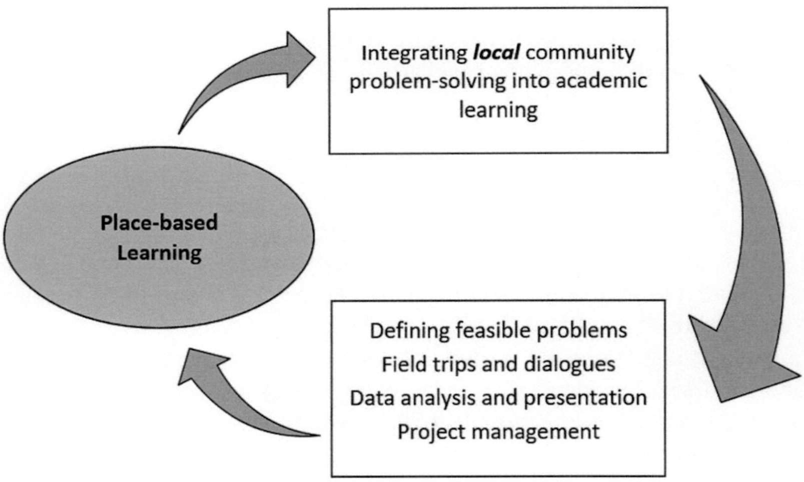

Fig. 5 Main components of the place-based learning process

community improvement. But despite being locally relevant, the learning outcome is designed to provide a basis for understanding important issues and problems in the global community of which the local community forms a part.

The place-based learning process is structured by the instructor since many issues such as public property, private access, and potential university ethics approval may be necessary. Usually the problem is defined by local knowledge and familiarity with the neighborhood or by simply walking around and observing the possibilities for researchable questions (Hupy et al. 2005). The formal investigation of the problem then proceeds by problem-based learning, inquiry-based learning, project-based learning, or service-based learning approaches.

Table 1 Summary of Book Chapters and their Pedagogies

Chapters	Lecture-based learning	Active learning	Problem- and inquiry-based learning	Project-based learning	Service learning	Place-based learning
Chapter 1	☑	☑	☑	☑	☑	☑
Chapter 2	☑				☑	☑
Chapter 3	☑	☑	☑	☑		☑
Chapter 4				☑	☑	☑
Chapter 5	☑			☑		
Chapter 6	☑			☑		
Chapter 7	☑			☑		
Chapter 8	☑		☑	☑	☑	☑
Chapter 9	☑				☑	☑
Chapter 10	☑	☑	☑	☑	☑	☑

Collectively, the teaching and learning pedagogies presented above provide higher education GIScience instructors with an overview of multiple pedagogies that can be used in various combinations to effectively engage students in a flexible learning process.

Book Chapters and Dominant Pedagogies

The chapters in this book form an impressive collection of innovative education research and case studies. Together, they cover one or more of the teaching and learning pedagogies identified. Table 1 provides a summary of where each chapter fits into the pedagogies described in the previous sections of this chapter. A first way to view the collection of chapters in this book is as case studies on how to implement specific pedagogies in a GIS context. A second way is to look at them as a foundation on which further improvements can be made incorporating new pedagogies to further motive and engage GIS students in a flexible learning process.

References

Baker, T. R., Battersby, S., Bednarz, S. W., Bodzin, A. M., Kolvoord, B., Moore, S., et al. (2015). A research agenda for geospatial technologies and learning. *Journal of Geography, 114*(3), 118–130.

Balram, S., & Dragicevic, S. (2005). An embedded collaborative systems model for implementing ICT-based multimedia cartography teaching and learning. In S. Mishra & R. C. Sharma (Eds.), *Interactive multimedia in education and training* (pp. 306–326). Hershey: Idea Group Publishing.

Balram, S., & Dragicevic, S. (2008). Collaborative spaces for GIS-based multimedia cartography in blended environments. *Computers & Education, 50*(1), 371–385.

Burrows, J., Bodzin, A., Anastasio, D., Sahagian, D., Bressler, D., Cirucci, L., et al. (2013). Using web GIS to enhance tectonics learning and geospatial thinking. *Science Scope, 37*(4), 29–37.
Clark, A. M., Monk, J., & Yool, S. R. (2007). GIS pedagogy, web-based learning and student achievement. *Journal of Geography in Higher Education, 31*(2), 225–239.
Doerr, N. M. (2013). Do 'global citizens' need the parochial cultural other? Discourse of immersion in study abroad and learning-by-doing. *Compare: A Journal of Comparative and International Education, 43*(2), 224–243.
Drennon, C. (2005). Teaching geographic information systems in a problem-based learning environment. *Journal of Geography in Higher Education, 29*(3), 385–402.
Elwood, S., & Wilson, M. (2017). Critical GIS pedagogies beyond 'Week 10: Ethics'. *International Journal of Geographical Information Science, 31*(10), 2098–2116.
Gordon, E., Elwood, S., & Mitchell, K. (2016). Critical spatial learning: Participatory mapping, spatial histories, and youth civic engagement. *Children's Geographies, 14*(5), 558–572.
Hupy, J. P., Aldrich, S. P., Schaetzl, R. J., Varnakovida, P., Arima, E. Y., Bookout, J. R., et al. (2005). Mapping soils, vegetation, and landforms: An integrative physical geography field experience. *The Professional Geographer, 57*(3), 438–451.
King, E. (2008). Can PBL-GIS work online. *Journal of Geography, 107*(2), 43–51.
Kinniburgh, J. (2010). A constructivist approach to using GIS in the New Zealand classroom. *New Zealand Geographer, 66*(1), 74–84.
Solari, O. M., & Schee, J. (Eds.). (2015). *Geospatial technologies and geographic education in a changing world: Geospatial practices and leassons learned*. Tokyo: Springer.
Tate, N. J., & Jarvis, C. H. (2017). Changing the face of GIS education with communities of practice. *Journal of Geography in Higher Education, 41*(3), 327–340.
Unwin, D. J., Foote, K. E., Tate, N. J., & DiBiase, D. (Eds.). (2011). *Teaching geographic information science and technology in higher education*. Chicester: Wiley.

The Evolution and Definition of Geospatial Literacy

Lynn Moorman

Abstract In all areas of GIScience, a fundamental premise of creating, using, and presenting geospatial data is that we understand how to portray, and extract information from representations of, that data. The ability to relay and extract information from representation (usually text) is termed basic literacy and, when put in the context of geospatial information representation, is considered geospatial literacy. Literacy of printed language is a fundamental curricular outcome of education at all levels. Recently however, the scope of literacy has expanded to encompass more than linear text, as visuals and graphics are also recognized as conveying meaning and are considered valid communication modes within a society. As geospatial technologies, data, and representations increasingly emerge onto the educational landscape, the need for relevant assessments, educational resources, professional development, and curricula to support effective geospatial literacy teaching and learning grows. Geospatial literacy is required to make meaningful products and extract meaningful information from geospatial data to inform and support our decisions. Geoliteracy is a term that has been used to describe the basic geographic declarative knowledge of a population, an ability to make geographic decisions, geographic reasoning, and a general understanding of geographic concepts. However, a formal definition of geoliteracy grounded in literacy and educational research and theory is more specific and critical to inform the necessary supports for improving and assessing geospatial literacy, both for teaching and learning. A description of the meaning of literacy, the ways in which it is measured globally and within Canada, and how that applies to our conception of literacy in the geospatial disciplines is provided here.

Keywords Geoliteracy · Geospatial · Literacy · Geography · Education · Representation · Geographic knowledge

---

L. Moorman (✉)
Mount Royal University, Calgary, Canada
e-mail: lmoorman@mtroyal.ca

S. Balram, J. Boxall (eds.), *GIScience Teaching and Learning Perspectives*, Advances in Geographic Information Science,
https://doi.org/10.1007/978-3-030-06058-9_2

Introduction

Geoliteracy is a popular term in the geography vernacular. National Geographic has committed to improving geoliteracy of citizens, particularly since reports of "geographic illiteracy" arose in the media to describe the dismal results of surveys of young American adults' geographic knowledge (Roper 2002, 2006). The media's take on these surveys were a double-edged sword to geographers. On the one hand, the focus on the need for geographic education was positive for the discipline, but on the other hand, the nature of the surveys, focusing on declarative knowledge, perpetuated the public view of geography as a subject dominated by memorization of names and locations, and geoliteracy as the ability to recall information. To address this and clarify the definition, National Geographic described geoliteracy as "an ability to reason about earth systems and interconnections" that is required in order to make informed decisions (National Geographic Society 2014). This definition is still vague, especially for educators trying to build and assess geoliteracy in their classrooms. Ironically, literacy – traditionally considered as the ability to read and write – is well defined and is a primary goal of education around the world. The term literacy does not only refer to memorization of declarative facts, or about how we make decisions, it is about reading and writing representations of ideas and knowledge, whether in text or graphics, to enable the acquisition of new knowledge, which ultimately can support our decision-making. Therefore, it seems evident that any definition of the term geoliteracy must be related and grounded in the understanding of the definition of literacy itself.

The term *geoliteracy* is rarely used in the academic literature, and when it is, it is *described*, not *defined*, in an academic sense (Carano and Berson 2007; Guertin et al. 2012). Canadian Geographic Education uses the term *geographic literacy* within their mandate and objectives, describing it in a broader sense of knowledge and skills (Canadian Geographic Education 2014). Beyond geographic literacy, they also define the term geographic *fluency* as "preparation sufficient for successful postsecondary study in subjects that require geographic skills and understanding" (para. 5). *Geospatial literacy* is the more accepted academic term, defined in a manner more aligned with working definitions of literacy (Maclachlan et al. 2014; Moorman 2014). For this reason, geoliteracy is referred to as geospatial literacy in this chapter.

The need to clarify the definition of geospatial literacy is rooted in the increased recognition of the importance of spatial thinking within the STEM or STEAM disciplines (science, technology, engineering, art, math) and the ability to provide a perspective capable of addressing large and complex problems (Wai et al. 2009). Explicating geospatial thinking in terms that can be made teachable and assessable is where geospatial literacy plays a role. The practice and assessment of literacy form the cornerstone of North American education systems, aligned with the recognition that increased literacy improves knowledge and understanding. In the language arts, teaching and assessing literacy through the reading and writing of text is a critical pathway toward the goal of students building and representing their understanding of written text in order to be contributing members of society.

Reciprocity is the key – as students build literacy skills, they are able to increase their knowledge, which helps improve skills and ultimately literacy. This is why literacy is often associated with freedom and emancipation, as expressed through the works of Paulo Freire and echoed by the United Nations:

> Indeed, literacy itself takes many forms: on paper, on the computer screen, on TV, on posters and signs. Those who use literacy take it for granted – but those who cannot use it are excluded from much communication in today's world. Indeed, it is the excluded who can best appreciate the notion of "literacy as freedom." (UNESCO 2003)

Literacy is considered to be so critical an indicator of social well-being and economic development that it is measured and assessed on a regular basis through the world. Literacy is considered throughout the educational spectrum, as it is introduced and nurtured early and is developed lifelong. Similarly, the conversation about geospatial literacy needs to consider K-12 education as much as postsecondary, for it is within K-12 classrooms that the foundational literacy elements are taught and practiced. In many societies around the world, the twenty-first century is exploding with access to different modes of communication and types of information, including spatial, and specifically, geographic, resulting in a constant reinvention of what it means to be literate. Spatial literacy is increasingly valuable, as "spatial modes of communication are extremely powerful" and are used to represent ideas and information across society (Dodge and Kitchin 2001, p. 2). The ability to use, analyze, and interpret images and maps is becoming more and more important in many scientific and industrial fields. In addition, some contend that the ability to use images and spatial technologies intelligently and critically is becoming a requirement to participate effectively as a citizen in modern society (Bednarz et al. 2006). This is recognized in learning policy documents, such as the Government of Alberta (2013) report which states that literacy informed by societal context is considered the keystone of learning:

> In the 21st century, literacy is more than reading and writing. Today and in the future, learners must develop expertise with a range of literacy skills and strategies to acquire, create, connect, and communicate meaning in an ever-expanding variety of contexts. (para. 2)

Spatial Literacy

Much of the geospatial literacy literature refers to broader spatial literacy and spatial thinking, which encompasses the geographic context, among others (Golledge et al. 2008; Goodchild 2001; Montello 1993). The term *spatial* also needs to be defined in order to situate the notion of geospatial. Spatial refers to how something (or numerous things) relates to or exists in space (Wade and Sommer 2006). Relevant spatial descriptors of an object include its location, relative position, orientation, or relationship to another object. For example, there are phenomena or concepts that may not be tied to geographic space but can be represented in a visual or graphic (spatial) framework, with positions and links between them – characters in a novel, genealogical history, or related ideas. Bednarz and Bednarz (2008) and

Sinton (2014) described the three types of contexts of space related to spatial thinking as proposed by the National Research Council seminal publication *Learning to Think Spatially* (2006):

1. Thinking *in* space (geography of life spaces). This concerns our physical environment and phenomena within and considers distribution, relation, and interaction of these phenomena, at or near the earth's surface. This includes building and navigating – generally our everyday activities involving our movement in, or utilization of, space.
2. Thinking ABOUT *about* space (geography of physical spaces). This context refers to understanding space and time as a framework upon which phenomena exist. An example is structure at many different scales – solar system or molecules, for example.
3. Thinking WITH *with* space (geography of intellectual spaces). This context brings a spatial dimension to abstract ideas. Thoughts and objects might not have a specific location nor be referenced to Earth, but they can be organized as meaningful relationships and logical sequences in a spatial framework. Concept maps and flow charts are examples. The inherent spatial idea, and the power, behind spatializing nonspatial information is that closer things are more similar than distant things (Goodchild 2011; Tobler 1970). Therefore in using space as a framework for ideas, we place, link, and connect related ideas closer together than more dissimilar ideas. This type of mapping explicates and makes visual the spatial metaphors used to describe relationships (close, distant, etc.).

The life spaces (thinking in space) and geographic spaces (thinking about space) are where our traditional and intuitive spatial thinking occurs. There are four scales of spatial thinking that are relevant to these contexts (Montello 1993).

1. Micro or body scale: referring to the smallest considerations of space, from nanotechnology to the human body
2. Figural: the vicinity within reach of the body
3. Environmental: the space of everyday bodily activity and/or an area that can be visually perceived
4. Geographic scale: large area and/or places that cannot be viewed from a single earthly viewpoint

The term *geographic* is inherently spatial and refers to how things exist relative to a very specific space and scale that of Earth (Wade and Sommer 2006). As such, the spatial scales that are relevant to geographers, as described above, are environmental and geographic (Golledge et al. 2008; Montello 1993). Within those scales, *geographic* refers to the exploration of spatial characteristics of place and people.

Goodchild (2001) suggested that it is important to distinguish the scale and subset of spatial that is actually geographic by using the term geospatial. Geospatial confers spatial theory within the context and principles of the geographic discipline and creates a critical niche. Spatial thinking in the geographic and environmental scales is referred to as geospatial thinking to identify the spatial subset and scale that is being considered (Golledge et al. 2008). Ishikawa's (2013) research suggested that geospatial thinking is not purely a subset of spatial thinking, but it is comprised of a range of elements not always consistent with spatial ability and therefore has merit as a unique way of thinking. Likewise, geospatial literacy is a subset of spatial literacy and yet may present its own unique characteristics.

As the root term *literacy* is evolving, so is the definition of spatial literacy. Developing an accepted definition of spatial literacy and assessment practices is a focus of much current research. Johnson (2008, p. 422) suggested, based on the definition of literacy by UNESCO, that spatial literacy is the ability to "understand the concept of space; apply processes of reasoning employing appropriate tools to determine spatial relationships between people, places or objects; and visualize or communicate those spatial relationships in various contexts." However, assessment of spatial literacy performance and thinking is not straightforward:

> We have a good understanding of what it means to be articulate and literate in the verbal domain. We can assess performance in spoken and written forms, and we focus on the teaching and learning of verbal thinking. By contrast, there is as yet no clear consensus about spatial thinking and, therefore, spatial literacy. (National Research Council 2006, p. 26)

There are, however, accepted general characteristics and practices that define such a spatially literate individual.

1. They have a habit of mind of thinking spatially: they know where, when, how, and why to think spatially.
2. They practice spatial thinking in an informed way: they have a broad and deep knowledge of spatial concepts and spatial representations, a command over spatial reasoning using a variety of spatial ways of thinking and acting, and well-developed spatial capabilities for using supporting tools and technologies.
3. They adopt a critical stance to spatial thinking: they can evaluate the quality of spatial data based on its source and its likely accuracy and reliability; can use spatial data to construct, articulate, and defend a line of reasoning or point of view in solving problems and answering questions; and can evaluate the validity of arguments based on spatial information (National Research Council 2006, p. 20).

Constituents of spatial literacy are proficiencies in knowledge of spatial concepts, spatial ways of thinking and acting, and spatial capabilities (National Research Council 2006, p. 18). As Johnson (2008) states, "spatial thinking uses the skills and competencies one must possess to be deemed spatially literate" (p. 423). This seems to pose a circular argument suggesting spatial literacy is required for spatial thinking and vice versa. This may be true – as in written literacy, where thinking and acting support one another. While language representation forms a framework for the consumption and representation of new conceptual knowledge, so may spatial representation. While babies quickly develop a spatial awareness, and think and act spatially, they are not yet considered literate. As they build concepts and expression, such as language or drawing, and are exposed to representations that relate to space and become literate, their conceptual understanding of space develops further, and they may be able to develop new ways of critically assessing and representing their spatial knowledge (Piaget and Inhelder 1956). "The ability to make sense out of forms of representation is not merely a way of securing meaning – as important as that may be – it is also a way of developing cognitive skills" (Eisner 1998, p. 8). This ties into Wiegand's (2006) description of reciprocity that builds when children are exposed to new representations of their existing knowledge, enabling new ways of considering and expressing what they learn in future. While Uttal (2000) has examined this phenomenon with paper maps, how this reciprocity happens with regard to building spatial understandings with geospatial technologies is only starting to be explored (Goodchild 2008; Uttal 2000; Moorman 2014).

The Role of GIS

In the K-12 environment, potentially the biggest disruptor of traditional geographic instruction and learning, and potential mobilizer of geospatial literacy, is geographic information systems (GIS). Early efforts to increase GIS adoption in education resulted in the value of explicit spatial thinking and the critical importance of geospatial literacy being recognized.

GIS has long been recognized as a powerful tool to facilitate geographic inquiry-based learning (Alibrandi 2003; Baker 2002; National Research Council 2006). As a tool to facilitate inquiry and authentic problem solving with real data, the infusion of GIS in schools was always promising. Additionally, there was the recognition that the functions and spatial questions addressed with a GIS were similar to spatial cognitive operations used in everyday life. Golledge (2003, p. 247) stated, "An implicit assumption of the GIS is that most of the procedures involved in its use represent fundamental components of naïve spatial experience or what might be called 'commonsense spatial knowledge.'"

Despite the promise of GIS for educational use and the actual growth of GIS in postsecondary education across the United States and Canada, the reality of K-12 classroom adoption was initially slow to realize (White 2005). Slow infusion of GIS into classrooms was rooted in a myriad of reasons, including hardware requirements, software and lab maintenance not aligned with K-12 resources, software interface complexity, time required for training, and access to relevant data (Kerski 2008). Moreover, an underlying issue identified was the low value placed on geographic thinking and spatial literacy, the low comfort level of students and teachers working with it, and lack of skill to be able to draw meaning from the spatial data (Baker 2002).

The original research question for the committee working on National Research Council's (2006) *Learning to Think Spatially* report was to investigate best pedagogical practice for utilizing GIS in K-12 education in order to support its adoption. However, the committee recognized that prior to addressing the question of GIS, they needed to come to an understanding and explanation of the value of spatial thinking underpinning the use of GIS. The committee came to a conclusion that spatial thinking was foundational yet undervalued and subsequently not sufficiently addressed in instruction or curricula. The need to foster spatial literacy was brought forward as an outcome of their report. Researchers were also facing similar situations, trying to measure the value of the use of GIS in a classroom, but limited by students and teachers being ill-prepared to understand and manipulate spatial data entering and exiting the GIS (Baker 2002; Meyer et al. 1999; National Research Council 2006). The sophisticated tasks enabled by the GIS tended to be beyond what many educators were familiar with. The concern of limited geospatial thinking elements evident in the curricula and apparent in learners was expressed as early as 1994 (Hearnshaw and Unwin 1994) and is best exemplified in statements made by Baker and White (2003) about research involving Grade 8 students in an attempt to incorporate GIS into an inquiry-based lesson.

> Students in this study did not appear to be sufficiently capable of creating generalizations across a series of data-points, engaging in basic pattern seeking, and explanatory activities …. Students were rapidly able to generate maps indicating where sampling events occurred, but they were unable to make generalized statements about the trends in the data…no students were adequately prepared to fully leverage the data set… . Teachers and students exhibited difficulty in terms of considering data spatially. (Baker and White 2003, p. 251)

Baker (2002) suggested the use of sophisticated geospatial technology in the classroom would not be meaningful until teachers and students had the disposition to think spatially and have a vocabulary to describe spatial terms. In Ratinen and Keinonen's (2011) study, they found while Google Earth was beneficial for preservice teachers' geographical thinking, their participants encountered difficulty in interpreting and analyzing map data within the program. These researchers' concerns are directly related to literacy – reading and understanding geospatial data. Forty percent of Ratinen and Keinonen's (2011) participants self-identified as having insufficient map skills. Westgard (2010) found that students in her study had difficulty with representation, with a mean score of 42% on representation-related questions. Tesar (2010) also mentioned the K-12 participants in her study self-identified as being not geographically literate. While their teachers were more generous in their assessment of the students in pre-intervention interviews, one teacher noted in her post-interview that while she initially considered her students to be geographically literate, the use of Google Earth in the classroom revealed the lack of geographic knowledge of her students, and she subsequently changed her assessment (Tesar 2010).

Research does indicate increased engagement, better contextual understanding, and even motivation to learn when geospatial technologies (GST) are used thoughtfully in the classroom (Alibrandi and Goldstein 2015). Improved spatial thinking and performance at different age levels are indicated anecdotally and through research (Kim and Bednarz 2013; Kolvoord, Charles, Meadow and Uttal 2012), however, research around how geospatial technology (GIS or virtual globes) increases students' spatial thinking skills or spatial literacy is still wanting . There is ambiguity around the definitions of spatial or geospatial literacy and geographic thinking and subsequent difficulty in measuring and assessing geospatial literacy practices. Ratinen and Keinonen (2011) claimed that the use of Google Earth increased participants' geographical thinking; however, this was based on participants' self-assessment, and it was not clear what their working definition of geographic thinking was nor what elements of geographic thinking they were specifically considering. Spatial thinking instruments have recently been developed to address the need to explicate what we mean by *thinking* in assessable terms (Kim and Bednarz 2013; Huynh and Sharpe 2013; Lee and Bednarz 2009). Situating geospatial literacy in the literacy landscape will further explicate elements that can then be addressed through pedagogy and assessment.

Literacy

Literacy is a term that seems at once easily understood, yet there is tension publicly and in the literature about its broadening definition. It appears that the term literacy is becoming a suffix tied to many different concepts (e.g., physical literacy, health literacy, and technology literacy). Therefore, it is crucial to clarify the definition of the term on its own, before extending that meaning in a discipline specific way. Literacy is the ability of an individual to make and share meaning through representation. This ability includes constructing, analyzing, and communicating knowledge in a physical form, whether through written language and technology or, as relevant to this topic, in a spatial manner. Literacy has traditionally and popularly been described as the ability of an individual or population to read and write printed texts. While that definition still exists today, it has also been challenged to accommodate the multiple means of sharing information and to represent the diversity of skills that an individual needs in order to successfully participate in society. Eisner (1998) described literacy beyond the typical definition relating primarily to written language, suggesting it is "the ability to decode or encode meaning in any of the social forms through which meaning is conveyed" (p. 9). Of critical importance today is our interpretation of what is *text* and recognizing that literacy is contextual, dependent on the tools and values in the society in which it is being considered. As the definition of literacy is evolving, there are new types of literacies being acknowledged and varied means of measuring literacy. This is critical to understand the context in which geospatial literacy is defined, the relevance of it across the spectrum of literacies, and the challenge of situating it in the literacy research landscape.

Literacy was once the privilege of well-educated elite of society, and literacy, in its most basic definition of being able to read and write in the dominant language of one's society, was not a common goal in Western cultures until the mid-nineteenth century. Printed information, texts, and maps were considered for centuries as instruments of power not for public consumption, with access guarded and restricted to political leaders and entourage (Monmonier 1996). The rise of literacy as an educational goal occurred with shift from individual education focus to mass schooling, and literacy was considered the means to achieving and measuring success of the mass schooling movement, as well as empowering the society as a whole (Graff 1991).

The meaning of literacy depends partially on the accepted definition of *text*. In the traditional view of literacy, the definition of text is taken up narrowly, of only

Table 1 Types of text recognized in the context of literacy by OECD

Continuous text	Noncontinuous text
Narration	Charts and graphs
Exposition	Tables
Description	Diagrams
Argumentation	Maps
Instructions	Forms
Documents/records	Information sheets
Hypertext	Calls and advertisements
	Vouchers
	Certificates

Continuous text refers to printed narrative (OECD 2010)

the printed word formed by letters. Kress (2003) considers that text must be made visible – it "must appear somewhere" but that texts could be semiotic – signs not necessarily tied to the sound of the letters. The Province of Ontario report, Literacy for Learning (2004) states that text is "a representation of ideas that can be shared over distance and time" (p. 6), giving the example of a map as a type of visual-graphic text. The Organization for Economic Cooperation and Development (OECD) recognizes the forms in Table 1 as text, either continuous or noncontinuous (OECD 2010; Canadian Council on Learning 2012a).

In exploring the similarities and differences between how users react to varied text or representations, both the medium and mode of representation must be considered. Medium refers to how we engage with the representation, through analogue book or digital screen, for example (Eisner 2008; Kress 2003; McLuhan and Fiore 1967). The mode is a "…culturally and socially fashioned resource for representation and communication" or *how* we read information, such as text or images (Kress 2003, p. 45).

The medium provides the platform with which one engages in meaning-making, and it can be tremendously influential in mediating and negotiating understandings (McLuhan and Fiore 1967). For the context of geospatial literacy, examples include viewing geographic representations on paper versus dynamic digital interfaces. McLuhan dominated early rhetoric around digital media with observations and terminology, such as media itself, which are relevant to the consideration of technology half a century later. He argued that media affects how people think, which has both individual and societal implications. The dynamic nature of the digital medium in geospatial representation, particularly the ability in virtual globes to actively zoom and change scales with corresponding change in image resolution, transforms how the viewer interacts and reads the image (Moorman and Crichton 2018), markedly different than interacting with the linear static representation in atlases.

Kress (2003) suggested that knowledge changes ontological shape when it changes modes. Therefore, multimodality of representation has "deep epistemological effects" (p. 51) affecting how and what we learn. In the current environment of multimedia and multimodality, mindfulness around knowledge representation and, subsequently, the definition of literacy are paramount. How the written word is read is different than how graphics and maps are read. Textual narratives channel thought into a linear path, with the words placed in a certain order to best represent the desired meaning. In these fixed perspectives of text, one must read the words in order to make sense, so there is a temporal and sequential intent. Maps and other graphic representations convey ideas situated within multidimensional space and invite exploration along many axes. Images have a spatial imperative, in which meaning is made by relative location of objects, typically central versus marginal (Kress 2003). The image invites personal agency, where readers decide where they will look and why. This agency also brings their subjectivity to an image – readers make meaning relevant to their own life and experience. This move to a more spatial logic, as opposed to the temporal sequential logic of text changes readers' perspective accordingly (Kress 2003). Cazden et al. (1996) also suggested that spatial/geographical elements are important in intertextuality, the complex ways of meaning-making through multiple modes.

Multimodal texts expand the notion of literacy from mere literacy of singular text to multiliteracies to encompass the plurality of texts and modes found within the modern context of our multilinguistic, multicultural, and interconnected society. This includes typical forms of representation and modes of meaning-making, including visual, audio, and spatial (in the sense of environmental spaces, architectural spaces, ecosystemic meanings, as well as geographical meanings) (UNESCO 2005). While at one time seen as a binary state – one was literate or not – literacy is now considered a multifaceted continuum, with foundational and composite literacies (Bybee 1997). The notion of multiliteracies is evidenced in current educational policy documents in Canada (Maclachlan et al. 2014). In today's classroom environment, the mode for geospatial learning for students has shifted from a predominantly *art-like* cartographic depiction of earth typical of atlases and globes to a blend with *photo-like* satellite images of earth typical of a virtual globe, which require different skill sets for interpretation (Plester et al. 2002; Moorman and Crichton 2018) and therefore a new understanding of what it means to be geospatially literate.

The Alberta Government states that the "21st Century challenges us to rethink what being a fully literate person is" (Government of Alberta 2010, p. 1). The National Council for Teachers of English described that a fully literate twenty-first-century citizen needs multiliteracies and that those would need to be malleable. Implications for addressing these new literacies run across all facets of educational practice, from instructional design, facilitation, and assessment. Central to this idea is the learner's own awareness – from intentionality of the selection of tools they choose to use to the ethical behavior and choices demonstrated in a potentially global online audience for their actions and work (Government of Alberta 2010). The focus on the transaction of learning appears across the educational landscape.

The Government of Alberta (2010) also recognizes within the multiliteracies that some literacies are so fundamental they transcend the boundaries of academic disciplines (e.g., prose literacy), while others are more content specific in terms of knowledge, skills, and disposition (e.g., science literacy) or may be process specific (e.g., digital literacy). The Government of Alberta (2010) breaks the meaning of literacy into knowledge, skills, and dispositions, which aligns with other governmental policy definitions and also with spatial literacy definitions from the National Research Council (2006).

Literacy has been recognized as an indicator to the well-being of citizens (Canadian Council on Learning 2012a; Employment and Social Development Canada 2014; OECD 2010; UNESCO 2005). To this end, international surveys have been conducted to assess literacy levels. There have been three foundational or everyday literacies measured somewhat consistently – prose, document, and quantitative literacy/numeracy. Other literacies, such as scientific literacy, and health literacy, as well as problem solving, are also assessed. In these literacy assessment instruments, geospatial literacy is ironically all over the map. While map reading is not included in all literacy assessments, it is included in the 1994–2003 International Adult Literacy Surveys as a component of *document* literacy but, as of 2003, is also included in the *numeracy* assessment, according to Employment and Social Development Canada (2014). The major surveys conducted and their types of literacy measured are described in Table 2.

Literacy assessment results have spatial distributions and patterns, as they are reported according to political or administrative area. A frequent representation of literacy scores is a map, as shown in Fig. 1, the Canadian Council for Literacy's (CCL) results for literacy values across Canada. The values shown refer to document literacy,

Table 2 Definitions of literacies as used by official assessment instruments

International Adult Literacy Survey (IALS) 1994–1998	
24 countries	Age of participants: 16–65
Prose literacy	The knowledge and skills needed to understand and use information from texts, including editorials, news stories, poems, and fiction
Document literacy	The knowledge and skills required to locate and use information contained in various formats, such as tables, graphs, schedules, charts, forms, and *maps*
Quantitative literacy	The knowledge and skills required to apply arithmetic operations to numbers embedded on printed materials. These types of skills are used, for example, when balancing a checkbook, figuring out a tip, or completing an order form
International Adult Literacy and Life Skills Survey 2003 (ALL/IALSS) 7 countries	
Prose literacy	Same as IALS definition
Document literacy	Same as IALS definition
Numeracy (changed/evolved from IALS quantitative literacy)	A broader more inclusive measure of mathematics skills and conceptual mathematical knowledge. This expanded scale measures more than the ability to perform mathematical operations on numbers embedded in text by including many tasks that require no or little reading
Health literacy	*Derived* literacy. The skills to enable access, understanding, and use of information for health
PISA Programme for International Student Assessment 2000, 2003 (40 countries); 2006 (57 countries); 2009 (65 countries – 26,000,000) Canada 23,000 students, 1000 schools	

(continued)

Table 2 (continued)

Reading literacy	A range of tasks from retrieving specific information to demonstrating a broad understanding and interpreting different types of types and reflecting on its content and features: (i) form of reading material- continuous text (prose) and *noncontinuous text* (lists, forms, graphs, diagrams), (ii) type of reading task – proficiency in retrieving information reflecting on content and form of text, and (iii) use for which the text was constructed – recognizing context or situation of text
Mathematical literacy	Content: broad concepts – chance, change, *space*, and shape Process: math competencies – (i) simple computations, (ii) solving straightforward problems, (iii) mathematical thinking, posing own problems Situations in which mathematics is used – private to scientific
Scientific literacy	Concepts – content as applied to real life in science of life and health, or earth and environment, and in technology Processes – (i) acquire, interpret, and act on evidence, describing, explaining, and predicting scientific phenomena, (ii) understanding scientific investigation, and (iii) interpreting scientific evidence and conclusions Situations – the application of scientific knowledge and the use of scientific processes

Compiled from Canadian Council on Learning (2012a)

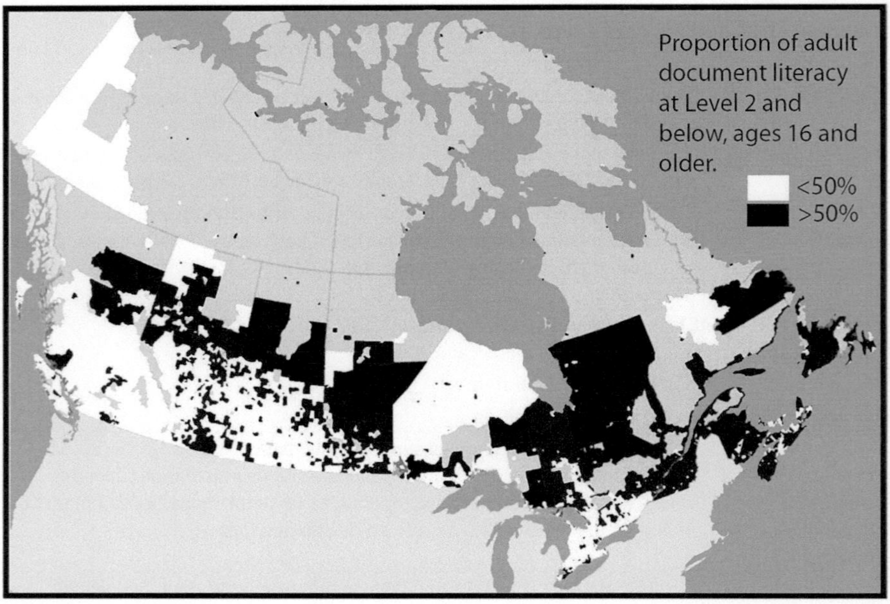

Fig. 1 The document literacy landscape across Canada originally compiled from 2006 Canadian census and 2003 International Adult Literacy and Life Skills Survey (IALLS) data, adapted from Canadian Council on Learning (2012b)

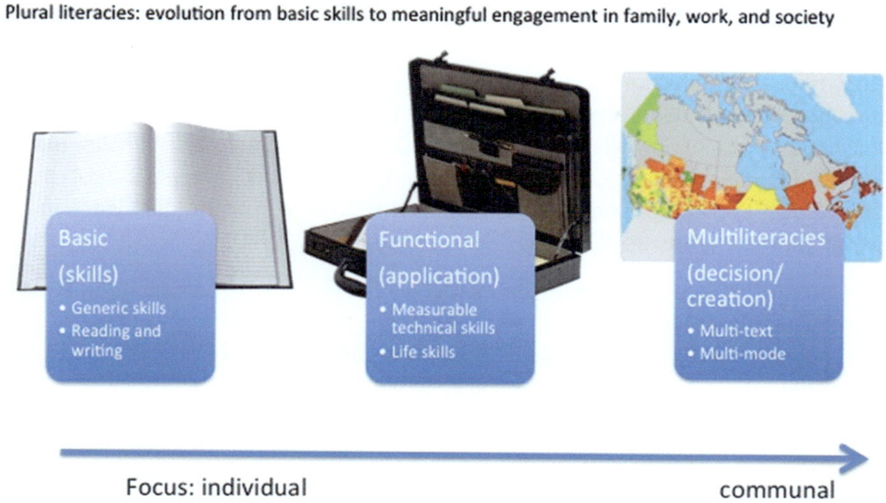

Fig. 2 Plural literacies represent literacies from personal to societal scales, evolving from basic skills to meaningful engagement in family, work, and society

which includes the literacy to read a map. Ironically, the data portrayed suggest less than half of the population in the black areas have the literacy to read the map.

The Canadian Council on Learning (2012a) and UNESCO (2004) described plural literacies as movement on the continuum of literacy from basic, to functional, to multiliteracies, as shown in Fig. 2, which relates to a progression from basic skills to competencies in applying those skills, to decision-making, and dispositions to create, which importantly require a shift in focus from the individual to the community. Gee (2012) suggested that literacy requires mastery of a second discourse outside of the home-based discourse in order to communicate as a society, thus "literacy is always plural" (p. 176). Collins and Blot (2003) stated that considering literacy as communication at only one scale and one mode (printed text), and not as a plurality, does a disservice to education, students, and society. As predicted by Alvermann and Hagood (2000), the notion of literacy is reinventing itself and continues to evolve. The solid foundation of literacy remains rooted, however, in three elements of literacy, as described by the Alberta Government (2010) and the National Research Council (2006), knowledge, skills, and dispositions.

Literacy Elements: Knowledge, Skills, and Dispositions in the Geospatial Context

Knowledge is the underlying foundation of what one knows about the world and is comprised of facts and understandings that are contextual (Calderhead 1996; Guerrero 2005). Knowledge is culturally influenced in both how it is gained and in how it is employed (Brown et al. 1989). As a construct, knowledge is characterized by connectivity, a "large integrated system difficult to isolate and dissect" (Guerrero 2005, p. 252). Knowledge can be thought of as subject or content-specific information, like geographic or scientific knowledge, and it can also relate to different domains or types of information. Examples from Shulman's (1986) work around teachers' types of knowledge included pedagogical knowledge, subject matter knowledge, and pedagogical content knowledge. Beyond this application-type of knowledge, knowledge comes in different forms. Three forms of knowledge in the geographic literature are (Golledge 2002; MacEachren 2004):

1. Core/declarative knowledge: the factual declarative core of information, knowledge about places
2. Content/configural knowledge: the understanding of concepts and processes and spatial relationships, knowledge between places
3. Procedural/functional: knowledge of how to do something

Core/Declarative Knowledge

Core or factual knowledge describes the type of explicit knowledge that can be codified, declared, and could be assessed as correct or incorrect. Core geographic knowledge can encompass declarative knowledge (Golledge 2002), descriptive knowledge (Madanes 1999), knowledge *of* space (Eliot 2000), and condition knowledge (Gersmehl and Gersmehl 2007). This type of knowledge typifies how geography is often referred to in mainstream society, as recall knowledge of the names and locations of countries and capital cities. While the perception of geography being a subject consisting only of rote memorization is frustrating for geographers, various types of geographic knowledge beyond core knowledge are now being recognized among educators. In the United Kingdom, core or geographic knowledge is recognized as an enabler of geographic thinking, but they strongly argue it does not constitute geographic thinking in and of itself. However, core knowledge of location constitutes most measures of geospatial literacy.

Much has been written recently about the role of knowledge and academic content in the learning process (Thomas and Seely Brown 2011). Thomas and Seely Brown (2011) identify that knowledge is essential in order to innovate. While core knowledge is viewed as a foundational piece of learning, it also must be contextual, experiential, or situated to provide the best conceptual understanding and not reside with the learner as a stand-alone piece of information (Wagner 1990). The United Kingdom's Geographical Association (2012) suggests core knowledge provides the context for understanding geographic processes and that learning principles and concepts without an underlying knowledge of location creates a meaningless or placeless concept "lacking in geographical authenticity," which is less likely to form strong schema of understanding (Geographical Association 2012, p. 2). Making meaning and being able to anchor declarative knowledge require tacit understanding and a personal connection to the information. Golledge described the changing nature of geographic knowledge in his Presidential Address to the Association of

American Geographers in 2002, from fact-based declarative knowledge to intellectual or creative knowledge as described by Eliot (2000).

Vocabulary can either be considered as declarative knowledge or its own type of knowledge (Compton et al. 2009). Vocabulary refers to the words and terminology used to describe knowledge. Discipline-specific vocabulary knowledge is recognized as a critical element in discipline concept understanding (Young 2005). Vocabulary knowledge is particularly important to those disciplines rich in technical terminology such as science, and geography may face a unique situation in that many different terms are used to describe a single spatial phenomenon in various disciplines and contexts (National Research Council 2006). Vocabulary is critical for representing knowledge in order to share information with others. Beyond representation, research has also indicated that spatial language plays a significant cognitive role in thinking about space, particularly understanding and communicating spatial relations (Kolvoord et al. 2012). Gentner et al. (2013) found that deaf children who had not been exposed to spatial language did not perform basic spatial nonlanguage tasks nearly as well as those with a spatial vocabulary, suggesting a relationship between spatial language and spatial performance. This finding agrees with Young (2005, p. 12) who identified vocabulary as "…the essential element of comprehending concepts in content areas." Many words used in science texts define concepts and that conceptual development is crucial to advancing learning in science. She added "students' level of understanding concerning their science vocabulary is an excellent predictor of their ability to understand science text" (Young 2005, p.12). Bednarz and Bednarz (2008) had similar conclusions in a geographic context, stating spatial vocabulary is critical to spatial thinking. In their findings, student success in using GIS to learn geography was heavily constrained by a lack of spatial vocabulary, a key component of literacy.

Building on the work of Nystuen (1963) and Papageorgiou (1969) among others, Golledge (1992, 2002) and Golledge et al. (2008) had been instrumental in determining theoretical spatial primitives and first- to fourth-order derivatives to provide a hierarchy and scope and sequence of geographic knowledge. Explicit attention to the vocabulary of primitives and then the associated derivatives provides the learner with a conceptual framework and vocabulary which enlarge their capacity for learning (Golledge et al. 2008).

Content/Configural Knowledge

Content, or configural knowledge, describes concepts, processes, and spatial relationships (Lambert 2011). Through configural knowledge, meaning can be made from core knowledge. It is what Gersmehl and Gersmehl (2007) describe as connection knowledge, which allows for comparisons and analogies to other places by connecting declarative knowledge. Survey knowledge of a space is considered configural knowledge, because the relationships between different features are known. Declarative knowledge of a location might consist of a latitude and longitude coordinate. Configural knowledge would place the location relative to other locations using a survey or map-like view.

Procedural Knowledge/Skills

Procedural knowledge consists of the general skills and specific practices that enable one to *do* geography and extract geographic information. With Google Earth, procedural knowledge is required for both interpreting the domain content (geographical features) and interacting with the technology. Procedural knowledge allows the declarative knowledge to be applied, and it is more flexible than static declarative knowledge (van Dijk et al. 1994). van Dijk et al. (1994) also suggest there are efficiencies in teaching and learning by providing equal focus on declarative and procedural knowledge.

Until recently, procedural knowledge in American geography education has been defined by skills, evident in the *Geography for Life* standards (Heffron and Downs 2012). The Canadian standards are closely aligned and also reflect this terminology. However, with the development of the US Road Map for Geographic Education (Bednarz et al. 2013), the focus shifted from the term *skill* to *practice*, which aligns with other disciplines (math, science) and acknowledges the complexity of thought and actions required in *doing* geography. Practices are considered to be goal-oriented and behavioral. In that sense they represent the embodied action and behaviors of doing an activity, with all of the declarative and tacit knowledge that entails.

The Road Map (Bednarz et al. 2013) identified six categories of practice, based on the five skills defined in the *Geography for Life* (Heffron and Downs 2012) standards, and added one additional. These categories were then condensed to give three broad types of geographic practice. Table 3 compares the skills in the US and Canadian standards (Canadian Council for Geographic Education 2001) with the new categories of practice. In the new nomenclature, skills become a subset of practice, much closer to the European geography education community's definition. In the British context, skill is a specific task that is engaged within a practice (United Kingdom 2013).

Structuring the elements of procedural knowledge from broadest to most specific provides a sense of the hierarchy of these terms in their current usage.

- Procedural knowledge (e.g., map the land cover of an area)
  - Categories of practice (e.g., acquire, organize, and analyze relevant geographic information)
    - Practices (e.g., identify data analysis strategies, find, and describe spatial patterns in the data)
      - Skills (e.g., use scale, employ interpretation elements to interpret image)

Geography for Life states that the content of geography is comprised of knowledge, skills, and perspectives (Heffron and Downs 2012). Perspective is the framework or lens in which problems and their potential solutions are viewed. Geographers employ both a spatial perspective (how things are located or related spatially or geographically) and an ecological perspective which recognizes the complex interconnectedness and interdependence of phenomena on Earth. These perspectives are important in the effective application or practice of knowledge and skills.

Table 3 Comparing geographic skills and practices from the Canadian and US Standards, the US Road Map, and the GER committee

Geography standards (Canada[a] and the United States[b]) [c]Skills	Road Map[d] General [c]Categories of practice	Road Map[d] Research [c]Categories of practice	Road Map[d] [c]Practices
Posing geographic questions	Posing geographic questions	Formulating geographic questions	(a) Identify problems or questions that can be addressed using geographic principles, models, and data; express problems and questions in geographic terms
Acquiring geographic information	Acquiring geographic information	Acquiring, organizing, and analyzing geographic information	(a) Identify geographic data that can help to answer a question or solve a problem (b) Collect data (including observations and measurements) about geographic phenomena, and/or gather existing data to help answer a question or solve a problem
Organizing geographic information	Organizing geographic information		(a) Organize data, and create representations of data to help solve a problem or answer a question
Analyzing geographic information	Analyzing geographic information		(a) Identify data analysis strategies that can be used to help solve a problem or answer a question (b) Find and describe spatial and temporal patterns in data, or find data that matches a pattern, to help solve a problem or answer a question (c) Construct an explanation or prediction for phenomena by comparing data to a model or theory
Answering questions and designing solutions	Answering questions and designing solutions	Explaining and communicating geographic patterns and processes	(a) Construct an answer to a question or a solution to a problem using geographic principles, models, and data (b) Evaluate one or more answers to a question or solutions to a problem using geographic principles, models, and data
—	Communicating geographic information		(a) Inform or persuade an audience using geographic principles, models, and data

[a]Canadian Council for Geographic Education (2001)
[b]Heffron and Downs (2012)
[c]Denotes the terminology used
[d]Bednarz et al. (2013)

Dispositions

Dispositions are a natural or acquired combination of qualities that a person demonstrates, sometimes referred to as elements of character. Dispositions can have a strong influence on student achievement (Costa and Kallick 2014), and some consider them to be a stronger influence than individual ability (Lange and Adler 1997 in Lai and Viering 2012). The recognition of *habits of mind* as a significant component of spatial literacy (National Research Council 2006) has further emphasized the need for a better understanding of what constitutes and influences disposition and affection in the geographical discipline.

Dispositions are tendencies to behave or perform in a certain way. In a learning context, dispositions refer to thinking habits, patterns of intellectual behavior, or how a learner approaches a problem (Tishman and Andrade 1997). The *disposition effect* refers to difference between what abilities someone has and what they are inclined to do. In the spatial thinking realm, a student may have mapping knowledge and skills, but not be predisposed to approach a problem in a spatial way. Therefore, their general disposition to not think spatially limits the application of their geography knowledge and skills. This idea also relates back to the messaging of the learner's own awareness and the intentionality of the tools they choose to solve a problem (Government of Alberta 2010). This understanding emphasizes the need to have all three elements of knowledge, skills, and dispositions working as a triad to enable geospatial literacy.

Dispositions are habitual (Case 2005) and can be predictable, but they are not necessarily automatic (Ennis 1996). Costa and Kallick (2014) describe dispositions as a "cluster of preferences, attitudes, and intention, plus a set of capabilities that allow the preferences to be realized in a particular way" (p. 19), and Facione et al. (1994) call them a "constellation of attitudes, intellectual virtues, and habits of mind" (p. 346).

The literature holds a wide range of examples and scales of what is a definable and distinct disposition (Tishman and Andrade 1996). As an example, overarching broad dispositions include mindfulness (Langer 1989; Salomon 1994), fair-mindedness (Paul 1990), and critical-spiritedness (Siegel 1999; Facione et al. 1994). A list of dispositions that foster an overall spatial thinking disposition appears to be a gap in the literature. However, work by Jo and Bednarz (2014) focused on dispositions to teach spatial thinking. Thinking dispositions are well described in the critical thinking literature. Critical thinking is defined by Ennis (1985) as "…reasonable reflective thinking focused on deciding what to believe or do" (p. 45) and identified by Case (2005) as a quality, not type, of thinking. As such, critical thinking may be the result of a number of dispositions working in harmony, rather than one disposition alone and onto itself (Balcaen, pers.comm). This definition of critical thinking as an indication of quality of thought, and the product of multiple habits of mind, situates critical thinking in the spectrum of literacy, directly relating basic literacy to improved decision making, as per the original National Geographic definition of geoliteracy.

> The ideal critical thinker is habitually inquisitive, well-informed, trustful of reason, open-minded, flexible, fair-minded in evaluation, honest in facing personal bias, prudent in making judgments, willing to reconsider, clear about issues, orderly in complex matters,

> diligent in seeking relevant information, reasonable in the selection of criteria, focused in inquiry and persistent in seeking results which are as precise as the subject and the circumstance of inquiry permit. (Facione 1990, p. 3)

A comparison of five lists of critical thinking dispositions, with identified commonalities, is indicated in Table 4. The list by Perkins et al. (1993) seems to most thoroughly encapsulate the common ideas. The dispositions of inquisitiveness, analyticity, and truth-seeking appear to tie into motivation to learn and identify a problem, while open-mindedness, systematicity, carefulness, and self-confidence reflect more on the process of learning and problem solving. Three dispositions that were recognized only by single authors were the following: (1) take into account the feelings and thoughts of other people (Ennis 1996), (2) interdependence (Costa 1991), and (3) maturity to be judicious in one's decision-making and recognizing and accepting ill-structured problems (Facione et al. 1994). In Moorman (2014), key dispositions that influenced geospatial literacy were creativity (open-mindedness), curiosity (inquisitiveness), analyticity, truth-seeking, and self-confidence, aligning well with the critical thinking dispositions articulated by Perkins et al. (1993) and Facione et al. (1994).

Geospatial Literacy: A Nutshell

Geospatial literacy involves reading and comprehending representations of Earth or phenomena on Earth. These representations "help us remember, understand, reason, and communicate about the properties of and relations between objects represented in space" (National Research Council 2006, p. 27). As per the general definition of literacy (Canadian Council on Learning 2012a; Government of Alberta 2009, 2010; National Research Council 2006), geospatial literacy can be deconstructed into components of geospatial knowledge, skills, and dispositions, recognizing interconnectedness among these.

With all of these considerations, geospatial literacy is about communicating effectively through geospatial representation. It is considered to be an amalgam of knowledge, skills, and dispositions required to communicate (write) and comprehend (read and analyze) representations of geospatial information. Maclachlan et al. (2014) provide a similar definition, "the ability to conceptualize, capture, and communicate spatial phenomena." Defining geospatial literacy with specific emphasis on knowledge, skills, and dispositions, rather than simply as an ability, facilitates the definition of learning outcomes that can be assessed. The US National Standards for Geography, as written in the *Geography for Life* document (Heffron and Downs 2012), are similarly constructed, such that geospatial literacy outcomes involving representation can easily be aligned with the standards. In fact, the first two standards relate specifically to reading, analyzing, and communicating with geographic representations, both analogue and mental maps. Likewise, the United Kingdom specifies the use of maps, GIS, and geographic data, and the making of maps and sketch maps, within their General Certificate of Secondary Education (GSCE) geography qualifications (United Kingdom 2013). Canada's current standards

Table 4 Comparison of critical thinking dispositions as reported in the literature

Perkins et al. (1993), Facione et al. (1994)	Costa (1991)	Ennis (1996)	Lai and Viering (2012)
1. *Open-mindedness: the disposition to be broad and adventurous* The tendency to be open-minded, to explore alternative views; an alertness to narrow thinking; the ability to generate multiple options	Flexibility	Open-minded: seriously consider other points of view and be willing to consider changing one's own position Withhold judgment when the evidence and reasons are sufficient to do so Look for alternatives	Open- or fair-mindedness Flexibility Respect for and willingness to entertain others' viewpoints
2. Inquisitiveness: *the disposition toward sustained intellectual curiosity* The tendency to wonder, probe, find problems, a zest for inquiry; an alertness for anomalies; the ability to observe closely and formulate questions		Seek and offer reasons	Inquisitiveness
3. Analyticity: *the disposition to clarify and seek understanding (build explanations and understandings)* A desire to understand clearly, to seek connections and explanations; an alertness to unclarity and need for focus; an ability to build conceptualizations		Seek and offer reasons Try to be well informed Be clear about the intended meaning of what is said, written, or otherwise communicated	Propensity to seek reason Desire to be well informed
4. Systematicity: *the disposition to be planful and strategic* The drive to set goals, to make and execute plans, to envision outcomes; alertness to lack of direction; the ability to formulate goals and plans.		Determine and maintain focus on the conclusion or question	
5. *The disposition to be intellectually careful* The urge for precision, organization, thoroughness; an alertness to possible error or inaccuracy; the ability to process information precisely	Craftsmanship	Be careful Seek as much precision as the situation requires Take the total situation into account	
6. Truth-seeking: *the disposition to seek and evaluate reasons* The tendency to question the given, to demand justification; an alertness to the need for evidence; the ability to weigh and assess reasons	Efficacy	Use one's critical thinking abilities	
7. Self-confidence: *the disposition be metacognitive* The tendency to be aware of and monitor the flow of one's own thinking; alertness to complex thinking situations; the ability to exercise control of mental processes and to be reflective	Awareness	Try to be reflectively aware of one's own beliefs	

identify map and globe skills, but do not mention digital representations nor are they subdivided into knowledge, skills, or dispositions. A recent effort to revise those standards may reflect a change in how geospatial literacy is considered.

Education Implications

The implications for education at all levels focus on the role of representation and the symbiotic nature of the three elements of geospatial literacy – knowledge, skills, and dispositions.

First, it is important to remember that forms of representation are not neutral nor are the tools we use (Eisner 2008). External representations affect what is cognitively consumed. As Senge explains, "we don't describe the world we see, we see the world we can describe." For example, information on a map and information on a virtual globe will each be taken up differently and affect knowledge construction in different ways (Papert 1980). Today's changing nature of information, from analogue to digital, and an increasing use of information visualization make it critical to look at both the media and mode of representations and how both inform (or perhaps misinform) what and how people learn. Eisner claimed that representations (e.g. maps) are not mere depictions of information – the end of a learning process, as often seems the case in education – they are in fact products of thought and cognitive artifacts that may be valuable tools for learning as they are made (Eisner 1998). Thus, they do not only capture knowledge and meaning, but they may be a means, in and of themselves, to develop cognitive skills and to understanding culture. This idea can be viewed in two ways: first that the medium one chooses to create with influences how one thinks, and second, there is learning involved in the creation of the representation, as suggested by Papert's idea of constructionism (Papert 1980). Kress (2003) recognized the reciprocal relationship between the book and writing – where the form one reads then informs what is produced. This same principle applies to cartography, argued Wiegand (2006), noting student sketch maps of the world are predominantly of the same general projection and orientation of common classroom Mercator maps that commonly privilege Western Europe at the center and expand the northern and southern latitudes disproportionally to equatorial areas. If representations are not neutral, it is important to be able to have exposure to, and the literacy to read, different types of representations in order to build a less biased construct of the world in which we live.

Eisner (2008), in writing about school curriculum, recognized the critical connection binding representation and literacy. He asked, "What forms of representation are emphasized? In what forms are students expected to become literate?" (p. 9). Repetitive use of the same types of geographic representations limits the scope of our construct of the world. It is important then to provide other means of representing the world and multi-scale space to allow students the greatest impartiality and flexibility in creating the constructs that are most accessible, meaningful, and sensible to them (Jacobson 2002). This is echoed by Liben and Downs (1989), who recommended viewing a variety of representations (scale, perspective, source data, projection) to dispel reified thinking. Assuming Kress (2003) and Eisner (1998) are correct, the reciprocal effect of these changes will shift how knowledge is subsequently represented. This invites new questions about representation in the classroom and how to teach and assess geospatial literacy. Using different types of representations demands attention to the literacy requirements of each, with a balance of knowledge, skills, and disposition. What is the affordance offered with each media/mode? What are the limitations? How are transformation (the processes of integrating and reorganizing knowledge within a single mode) and transduction (reconfiguration and shifting of knowledge between modes) accommodated and facilitated? These questions and the respective responses inform how geospatial representations and technologies are introduced into the classroom at all levels, the ways in which we can expect students to interact with them and express their own knowledge, and the geospatial literacy elements required.

Cartography is a useful example as it has experienced a paradigm shift in how its discipline of earth representation is understood, similar to the shift in educational theory from behaviorism to constructivism (MacEachren 2004). These shifts in perception were propelled by advances in psychology, specifically, in understanding how people learn. It is now understood that a map is not a vehicle for transmission of static information. Rather than being a passive receptacle for the cartographer's knowledge, the reader's own role is to engage with the information being presented by the cartographer and make sense of it based on his own prior knowledge and experience and actively construct new knowledge (MacEachren 2004). In this way, the purpose of maps has evolved from transmitters to enablers, and research has moved from modeling of symbols and map construction to user-map interaction and spatial cognition (MacEachren 2004; National Research Council 2006). "The user can combine map information with previous knowledge to produce conclusions that were not part of the initial map message" (MacEachren 2004, p. 9). Realizing that maps and earth representations do not necessarily convey the same meanings to all users suggests that research into the users themselves is required – user case studies, concerning cognitive and perceptual issues of geospatial representations – and thoughtful consideration of how geospatial literacy is assessed.

The second implication is that having geospatial literacy elements is not enough. Skills, knowledge, and dispositions work together, each supporting and enhancing the other in a symbiotic relationship. Learners need to have core and conceptual geographic knowledge, knowledge of how the technology works or how the representation was created, practices of extracting geospatial information, a means of representing and sharing that information, and an underlying disposition to consider problems in a geospatial context and choose the appropriate tools to address the problem in the first place. Having mapping skills is of no use if a person can't recognize a spatial question or does not have the desire to use them. This has a particular relevance to geospatial technologies in a twenty-first-century learning context, where geospatial data and technologies are accessible to students, and students may not rely on teachers for all of their technology instruction. As we are asking more of students in terms of metacognition – understanding their own interests, learning styles, and preferences – and promoting technology use, we should also be ensuring they have the elements of literacy to effectively use technology to support their learning. In a geospatial context, this includes having enough geospatial knowledge to inspire questions and want to know more, a disposition to seek out the most appropriate geospatial tools to address the questions, and the geospatial skills to use the technology in a discipline-relevant way. If one of the elements of knowledge, skills, or dispositions is lacking, the potential power of geography may be lost.

Conclusion

Literacy does not simply refer to understanding but how we negotiate understanding through representation, typically *text*. The notion and definition of text have expanded from printed text to encompass nonlinear forms, such as the modes of images and maps, and the enabling media of digital geography platforms and applications. Multimodal and multimedia geospatial representations are ubiquitous in our society, emphasizing the need for literacy in negotiating our understandings with them. These representations may provide an iterative, tandem learning opportunity if they enable someone to think spatially or to consider phenomena in a spatial context, in a way that connects their existing concepts to the new representation (Davies and Uttal 2007; Liben 2000). There is a sense of reciprocity, of a positive feedback loop, that builds when exposure to new representations of existing knowledge enables new ways of considering and expressing both existing and future knowledge (Wiegand 2006), highlighting the importance of learning with representations expressed in multiple modes and media.

Building on the definition and use of the term literacy in the Canadian educational context, geospatial literacy refers to the amalgam of knowledge, skills, and dispositions required to communicate and comprehend geospatial information through representation. Geospatial literacy is the key to turning geospatial data into information and then into personal knowledge that can be used to make decisions, to empower, to further question, and to communicate and report back to society. The three elements of geospatial literacy (knowledge, skills, and dispositions) were presented separately here; however, these elements are symbiotic and cannot effectively be considered in isolation. Being geospatially literate means that each of these elements informs the others in an iterative transaction of understanding and meaning-making.

An underlying issue for geospatial academic programs and industry itself is the same issue Baker (2002) faced in his research into GIS in the classroom – a lack of foundational geospatial literacy impedes the impact and effectiveness of using geospatial technology or applying geospatial data. In society in general, a lack of geospatial literacy reduces the effectiveness of the messaging that geographic products provide – information necessary for work, for social interactions, for travel, and critical in times of disaster.

Geospatial literacy elements are necessary at all ages but demand attention and practice and reinforcement. Research suggests that children do not make the same sense of imagery or maps as adults (National Research Council 2006). Therefore, it is important to consider geospatial representations and required geospatial literacy from multi-age perspectives, along with evidence of constraints or affordances, to maximize effective learning opportunities across the educational landscape. Supports for all levels of education, for geographers and non-geographers alike, are required to explicate appropriate geospatial literacy elements and provide tools for instruction, learning, and assessment.

References

Alibrandi M. & Goldstein D. (2015). Integrating GIS and Other Geospatial Technologies in Middle Schools. In: Muñiz Solari O., Demirci A., Schee J. (eds) Geospatial Technologies and Geography Education in a Changing World. Advances in Geographical and Environmental Sciences. Springer, Tokyo.

Alibrandi, M. (2003). *GIS in the classroom: Using geographic information systems in social studies and environmental science*. Portsmouth: Heinimann.

Alvermann, D., & Hagood, M. (2000). Critical media literacy: Research, theory, and practice in "new times". *The Journal of Educational Research, 93*(3), 13.

Baker, T. (2002). *The effects of geographic information system (GIS) technologies on students' attitudes,* self-*efficacy, and achievement in middle* school science *classrooms.* (Doctoral Dissertation, University of Kansas, Kansas).

Baker, T. R., & White, S. (2003). The effects of GIS on students' attitudes, self-efficacy, and achievement in middle school science classrooms. *Journal of Geography, 102*(6), 243–254.

Bednarz, S. W., Acheson, G., & Bednarz, R. S. (2006). Maps and map learning in social studies. *Social Education, 70*(7), 298–404.

Bednarz, R. S., & Bednarz, S. W. (2008). The importance of spatial thinking in an uncertain world. In D. Sui (Ed.), *Geospatial technologies and homeland security* (Vol. 94, pp. 315–330). Dordrecht: Springer.

Bednarz, S. W., Heffron, S., & Huynh, N. T. (Eds.). (2013). *A roadmap for 21st century geography education: Geography education research*. American Association of Geographers: Washington, DC.

Brown, J. S., Collins, A., & Duguid, P. (1989). Situated cognition and the culture of learning. *Educational Researcher, 18*(1), 32–42.

Bybee, R. (1997). Towards an understanding of scientific literacy. Scientific Literacy:Science Education and Secondary School Students. In W. Graber and C. Bolte (Eds.). Science Literacy: An International Symposium (pp. 37-68). Kiel: IPN.

Calderhead, J. (1996). Teachers; beliefs and knowledge. In D. C. Berliner & R. C. Calfee (Eds.), *Handbook of educational psychology* (pp. 709–725). New York: Macmillam.

Canadian Council for Geographic Education. (2001). *Canadian national standards for geography: A standards-based guide to K-12 geography*. Ottawa: The Royal Canadian Geographical Society.

Canadian Council on Learning. (2012a). *How do we measure literacy in Canada? What is* literacy *and how is it measured?* Retrieved Sept. 15, 2013, from http://www.ccl-cca.ca/CCL/Topic/Literacy/WhatisLiteracy.html#how

Canadian Council on Learning. (2012b). *Document literacy map*. Retrieved Sept. 15, 2013, from http://www.ccl-cca.ca/cclflash/documentliteracy/map_canada_e.html

Canadian Geographic Education. (2014). *Geographic literacy in Canada*. Retrieved Feb. 25, 2014 from http://www.cgeducation.ca/programs/geoliteracy/

Carano, K. T., & Berson, M. J. (2007). Breaking stereotypes: Constructing geographic literacy and cultural awareness through technology. *The Social Studies, 98*(2), 65–69.
Case, R. (2005). Moving critical thinking to the main stage. *Education Canada, 45*(2), 42–49.
Cazden, C., Cope, B., Fairclough, N., Gee, J., Kalantzis, M., Kress, G., et al. (1996). A pedagogy of multiliteracies for designing social futures. *Harvard Educational Review, 66*(1), 60–92.
Clagett, K. (2009). *Virtual globes as a platform for developing spatial literacy.* (Unpublished master's thesis, Universidade Nova de Lisboa, Portugal).
Collins, J., & Blot, R. (2003). *Literacy and literacies: Texts, power, and identity*. Cambridge: Cambridge University Press.
Compton, D., Elleman, A., Olinghouse, N., Lawrence, J., Bigelow, E., Gilbert, J., & Davis, G. N. (2009). The influence of in-text instruction on declarative knowledge and vocabulary learning in struggling readers. In R. Wagner, C. Schatschneider, & C. Phythian-Sence (Eds.), *Beyond decoding: The behavioural and biological foundations of reading comprehension*. New York: Guilford Press.
Costa, A. L. (1991). The school as a home for the mind. In A. L. Costa (Ed.), *Developing minds: A resource for teaching thinking*. Alexandria: ASCD.
Costa, A. L., & Kallick, B. (2014). *Dispositions: reframing teaching and learning*. Thousand Oaks: Sage.
Davies, C., & Uttal, D. (2007). Map use and the development of spatial cognition. In J. Plumert & J. Spencer (Eds.), *The emerging spatial mind* (pp. 219–247). Oxford: Oxford University Press.
van Dijk, H., van der Schee, J., Trimp, H., & van der Zijpp, T. (1994). Map skills and geographical knowledge. *International Journal of Geographical and Environmental Education, 3*(1), 68–80.
Dodge, M., & Kitchin, R. (2001). *Atlas of cyberspace*. Michigan: Addison-Wesley.
Eisner, E. (1998). *The kind of schools we need*. Portsmouth: Heinemann.
Eisner, E. (2008). The role of the arts in transforming consciousness. In *The brain and learning* (pp. 359–369). San Francisco: Jossey-Bass.
Eliot, J. (2000). *The nature and measurement of spatial intelligence*. College Park: Institute for Child Study, College of Education.
Employment and Social Development Canada. (2014). *Indicators of well-being in Canada*. Retrieved from http://www4.hrsdc.gc.ca/.3ndic.1t.4r@-eng.jsp?iid=79
Ennis, R. (1985). A logical basis for measuring critical thinking skills. *Educational Leadership, 43*(2), 44–48.
Ennis, R. (1996). Critical thinking dispositions: Their nature and assessability. *Informal Logic, 18*(2), 165–182.
Facione, P. A. (1990). *The California critical thinking skills test (CCTST)*. Millbrae: California Academic Press.
Facione, N. C., Facione, P., & Sanchez, C. A. (1994). Critical thinking disposition as a measure of competent clinical judgement: The development of the California Critical Thinking Disposition Inventory. *Journal of Nursing Education, 33*, 345–350.
Gee, J. P. (2012). *Social linguistics and literacies* (4th ed.). New York: Routledge.
Gentner, D., Özyürek, A., Gürcanli, O., & Goldin-Meadow, S. (2013). Spatial language facilitates spatial cognition: Evidence from children who lack language input. *Cognition, 127*, 218–240.
Geographical Association. (2012). *Thinking geographically*. From http://www.google.ca/url?sa=t&rct=j&q=&esrc=s&source=web&cd=4&ved=0CDUQFjAD&url=http%3A%2F%2Fwww.geography.org.uk%2Fdownload%2FGA_GINCConsultation12ThinkingGeographically.pdf&ei=xpqcU9y1EoL0oASmz4CoCg&usg=AFQjCNHrqrAv8axpYc7MDkIWBwzysQ3GEA&bvm=bv.68911936,d.cGU
Gersmehl, P. J., & Gersmehl, C. A. (2007). Spatial thinking by young children: Neurologic evidence for early development and educability. *Journal of Geography, 106*(5), 181–191. https://doi.org/10.1080/00221340701809108.
Golledge, R. G. (1992). *Do people understand spatial concepts: The case of first-order primatives*. Paper presented at the international conference GIS- from space to territory: Theories and methods of spatio-temporal reasoning, Pisa, Italy.

Golledge, R. G. (2002). The nature of geographic knowledge. *Annals of the Association of American Geographers, 92*(1), 1–14. https://doi.org/10.1111/1467-8306.00276.

Golledge, R. G. (2003). The Open Door of GIS. In R. Bechtel & A. Churchman (Eds.), Handbook of Environmental Psychology (p. 244-255). Hoboken, NJ: John Wiley & Sons.

Golledge, R. G., Marsh, M., & Battersby, S. (2008). Matching geospatial concepts with geographic educational needs. *Geographical Research, 46*(1), 85–98.

Goodchild, M. (2001). *A geographer looks at spatial information theory*. Paper presented at the international conference, COSIT, Morro Bay, CA.

Goodchild, M. (2008). The use cases of digital earth. *International Journal of Digital Earth, 1*(1), 31–42.

Goodchild, M. F. (2011). Spatial thinking and the GIS user interface. *Procedia – Social and Behavioral Sciences, 21*, 3–9. https://doi.org/10.1016/j.sbspro.2011.07.002.

Government of Alberta. (2009). *Living literacy: a framework for Alberta's next generation economy*. [Report] Edmonton, AB: Retrieved from http://aet.alberta.ca/community.aspx

Government of Alberta. (2010). *Literacy first: A plan for action*. [Report] Edmonton, AB.

Government of Alberta. (2013). *Literacy: A world to discover*. [Report] Retrieved from http://education.alberta.ca/teachers/program/literacy.aspx

Graff, H. (1991). *The literacy myth: Cultural integration and social structure in the nineteenth century*. New Brunswick: Transaction Publishers.

Guerrero, S. (2005). Teacher knowledge and a new domain of expertise: Pedagogical technology knowledge. *Journal of Educational Computing Research, 33*(3), 249–267.

Guertin, L., Stubbs, C., Millet, C., Lee, T.-K., & Bodek, M. (2012). Enhancing geographic and digital literacy with a student-generated course portfolio in Google Earth. *Journal of College Science Teaching*. National Science Teachers' Association, *42*(4).

Hearnshaw, H., & Unwin, D. (1994). *Visualization in geographical information systems*. Chichester: Wiley.

Heffron, S., & Downs, R. (Eds.). (2012). *Geography for life: National geography standards*. Washington, DC: Geography Education National Implementation Project.

Huynh, N. T., & Sharpe, B. (2013). An assessment instrument to measure geospatial thinking expertise. *Journal of Geography, 112*(1), 3–17. https://doi.org/10.1080/00221341.2012.682227.

Ishikawa, T. (2013). Geospatial thinking and spatial ability: An empirical examination of knowledge and reasoning in geographical science. *The Professional Geographer, 65*(4), 636–646.

Jacobson, R. D. (2002). *Representing spatial information through multimodal interfaces*. Proceedings sixth international conference on information visualization. July 10–12, 2002, London, England.

Jo, I., & Bednarz, S. (2014). Dispositions toward teaching spatial thinking through geography: Conceptualization and exemplar assessment. *The Journal of Geography, 133*(5), 198–207.

Johnson, A. (2008). Spatial literacy. In K. Kemp (Ed.), *Encyclopedia of geographic information science* (pp. 422–423). Thousand Oaks: Sage.

Kerski, J. J. (2008). The role of GIS in Digital Earth education. *International Journal of Digital Earth, 1*(4), 326–346. https://doi.org/10.1080/17538940802420879.

Kim, M. & Bednarz, R. (2013). Development of critical spatial thinking through GIS learning. Journal of Geography in Higher Education 37(3), 350-366.

Kolvoord, R., Charles, M., Meadow, N and Uttal, D. (2012). The impact of geospatial technologies in a dual-enrolment environment. Society for Information Technology and Teacher Education International Conference, Volume 2012.

Kress, G. (2003). *Literacy in the new media age*. New York: Routledge.

Lai, E., & Viering, M. (2012). *Assessing 21st century skills: Integrating research findings* (Pearson, Ed., p. 66). Vancouver: National Council on Measurement in Education.

Lambert, D. (2011). *The geography national curriculum: GA curriculum proposals and rationale* (p. 13). Sheffield: Geographical Association.

Langer, E. J. (1989). *Mindfulness*. Menlo Park: Addison-Wesley.

Lange, G. W., & Adler, F. (1997). Motivation and achievement in elementary children. Paper presented at the biennial meeting of the Society for Research in Child Development, April, Washington, D.C.

Lee, J., & Bednarz, R. (2009). Effect of GIS learning on spatial thinking. *Journal of Geography in Higher Education, 33*(2), 183–198. https://doi.org/10.1080/03098260802276714.
Liben, L. (2000). Map use and the development of spatial cognition: Seeing the bigger picture. *Developmental Science, 3*(3), 270–274.
Liben, L., & Downs, R. (1989). Understanding maps as symbols: The development of map concepts in children. *Advances in Child Development and Behavior, 22*, 145–201.
MacEachren, A. (2004). *How maps work*. New York: Guilford Press.
Maclachlan, J., Brodeur, J., Bagg, J., Chiappetta-Swanson, C., Vine, M., & Vajoczki, S. (2014). *An assessment of a blended learning model for information and geospatial literacy*. Toronto: Higher Education Quality Council of Ontario.
Madanes, R. (1999). *A cognitive account of teachers' descriptive knowledge*. (Doctoral Dissertation, University of California, Berkley).
McLuhan, M., & Fiore, Q. (1967). *The medium is the massage*. New York: Bantam Books.
Meyer, J. W., Butterick, J., Olkin, M., & Zack, G. (1999). GIS in the K-12 curriculum: A cautionary note. *The Professional Geographer, 51*(4), 571–578.
Monmonier, M. (1996). *How to lie with maps* (2nd ed.). Chicago: University of Chicago Press.
Montello, D. R. (1993). *Scale and Multiple psychologies of space*. Paper presented at the European conference COSIT, Marciana Marina, Elba Island, Italy.
Moorman, L. and Crichton, S. (2018). "Learner Requirements and Geospatial Literacy Challenges for Making Meaning with Google Earth," International Journal of Geospatial and Environmental Research, 5(3), Article 5.
Moorman, L. (2014). *Reciprocity of geoliteracy and representation: Children's learning with Google Earth*. (Unpublished PhD Dissertation, University of British Columbia, Okanagan).
National Geographic Society. (2014). *Geo-literacy*. Retrieved from http://education.nationalgeographic.com/education/geoliteracy/?ar_a=2
National Research Council. (2006). *Learning to think spatially*. Washington, DC: National Academies Press.
Nystuen, J. D. (1963). Identification of some fundamental concepts. *Michigan Academy of Science, Arts, and Letters, 48*(12), 373–384.
OECD. (2010). *PISA 2009 assessment framework key competencies on reading, mathematics, and science* (p. 296). Paris: OECD Publishing.
Ontario Education. (2004). Literacy for learning: The report of the expert panel on literacy in grades 4–6 in Ontario. Retrieved from www.edu.gov.on.ca/eng/document/reports/literacy/panel/literacy.pdf
Papageorgiou, G. J. (1969). Description of a basis necessary to the analysis of spatial systems. *Geographical Analysis, 1*(2), 213–215.
Papert, S. (1980). *Mindstorms*. New York: Basic Books Inc.
Paul, R. (1990). *Critical thinking: what every person needs to survive in a rapidly changing world*. Rohnert Park: Centre for Critical Thinking and Moral Critique, Sonoma State University.
Perkins, D., Jay, E., & Tishman, S. (1993). Beyond abilities: A dispositional theory of thinking. *Merrill-Palmer Quarterly: Journal of Developmental Psychology, 39*(1), 1–20.
Piaget, J., & Inhelder, B. (1956). *The child's conceptualization of space*. London: Routledge.
Plester, B., Richards, J., Blades, M., & Spencer, C. (2002). Young children's ability to use aerial photographs as maps. *Journal of Environmental Psychology, 22*, 29–48.
Ratinen, I., & Keinonen, T. (2011). Student-teachers' use of Google Earth in problem-based geology learning. *International Research in Geographical & Environmental Education, 20*(4), 345–358.
Roper, A. S. W. (2002). *National Geographic–Roper 2002 global geographic literacy survey*. Washington, DC: National Geographic Education Foundation. C510007671.
Roper, A. S. W. (2006). *National Geographic–Roper 2006 global geographic literacy survey*. Washington, DC: National Geographic Education Foundation. Retrieved from: http://www.nationalgeographic.com/roper2006/findings.html.
Salomon, G. (1994). *Interaction of media, cognition, and learning: An exploration of how symbolic forms cultivate mental skills and affect knowledge acquisition*. New York: Routledge.

Shulman, L. (1986). Those who understand: Knowledge growth in teaching. *Educational Researcher, 15*(2), 4–14.
Siegel, H. (1999). What (good) are thinking dispositions? *Educational Theory, 49*(2), 207–221.
Sinton, D. (2014) People's guide to thinking spatially.
Tesar, J. E. (2010). *The impact of a geographic information system on middle school students' geographic* literacy *and historical empathy.* (Dissertation/Thesis, Ohio University, Ohio).
Thomas, D. S. K., & Seely Brown, J. (2011). *A new culture of learning: Cultivating the imagination for a world of constant change*. Lexington: CreateSpace.
Tishman, S., & Andrade, A. (1997). *Critical squares: Games of critical thinking and understanding*. Englewood: Teacher Ideas Press.
Tishman, S. & Andrade, A. (1996) Thinking dispositions: A review of current theories, practices, and issues. Cambridge, MA: Harvard University, Available at: http://learnweb.harvard.edu/alps/thinking/docs/Dispositions.pdf (accessed January 15, 2014).
Tobler, W. (1970). A computer movie simulating urban growth in the Detroit region. *Economic Geography, 46*(2), 234–240.
UNESCO. (2003). *United Nations literacy decade*. Retrieved from: http://portal.unesco.org/education/en/ev.php-URL_ID=22420&URL_DO=DO_TOPIC&URL_SECTION=201.html
UNESCO. (2004). The plurality of literacy and its implications for policies and programmes. *UNESCO education sector Position paper*. Paris.
UNESCO. (2005). *Education for all: Global monitoring report: Literacy for life* (UNESCO Ed., Vol. 4, p. 445). Paris.
United Kingdom. (2013). *Geography GCSE subject content and assessment objectives*. Retrieved from http://www.education.gov.uk/schools/teachingandlearning/qualifications/gcses.
Uttal, D. (2000). Seeing the big picture: Map use and the development of spatial cognition. *Developmental Science, 3*(3), 247–286.
Wade, T., & Sommer, S. (2006). *A to Z GIS: An illustrated dictionary of geographic information systems*. Redlands: Esri Press.
Wagner, D. A. (1990). Who's a literate? Assessment issues in a global perspective. *Psychology & Developing Societies, 2*(1), 5–16.
Wai, J., Lubinski, D., & Benbow, C. P. (2009). Spatial ability for STEM domains: Aligning over fifty years of cumulative psychological knowledge solidifies its importance. *Journal of Educational Psychology, 101*, 817–835.
Westgard, K. S. W. (2010). *Google Earth in the middle school geography classroom: Its impact on spatial literacy and place geography understanding of students.* (Doctoral Dissertation, The University of North Dakota, ProQuest, UMI Dissertations Publishing).
White, S. (2005). *Geographic Information Systems (GIS) and instructional technology (IT) diffusion: K-12 student and educator conceptualizations.* (Unpublished PhD Dissertation, North Carolina State University).
Wiegand, P. (2006). *Learning and teaching with maps*. New York: Routledge.
Young, E. (2005). The language of science, the language of students: Bridging the gap with engaged learning vocabulary strategies. *Science Activities: Classroom Projects and Curriculum Ideas, 42*(2), 12–23.

Enabling Scientific Research Skills in Undergraduate Students During a Spatial Modeling Course

Suzana Dragićević and Taylor Anderson

Abstract Scientific research skills can be a valuable asset for undergraduate students pursuing spatial modeling and geographic information science courses. These skills provide students with a systematic means to think critically, solve complex geospatial problems, and contribute in meaningful ways to the scientific knowledge creation and dissemination process. In this study, targeted changes are described for an upper-division undergraduate spatial modeling course where students were guided through all the stages of the scientific research process. Students then developed geospatial solutions to real-world problems and communicated their results at a real scientific conference. Based on anonymous student feedback, the experience was perceived to be rewarding and the research skills gained to be a lifelong asset.

Keywords Spatial modeling · Geosimulation · GIS · Scientific research skills · Undergraduate students

Introduction

The undergraduate student research experience can have a strong impact on the future career choices of students (Trosset et al. 2008). Students' exposure to the scientific research process during a structured course with practical experience can provide multiple benefits. These benefits include enabling students to investigate problems in a systematic and standard manner, think critically using evidence-based decision-making, engage with current real societal problems, and be motivated for further graduate studies (Chen 1998). The scientific research process includes the stages of defining a problem, gathering evidence and quantitative data, conducting analysis, and communicating results to scientists and decision-makers (Ebenezer et al. 2010). Several studies have been conducted in North America (Lopatto 2007;

S. Dragićević (✉) · T. Anderson
Spatial Analysis and Modeling Research Laboratory, Department of Geography, Simon Fraser University, Burnaby, BC, Canada
e-mail: suzanad@sfu.ca; taylora@sfu.ca

S. Balram, J. Boxall (eds.), *GIScience Teaching and Learning Perspectives*, Advances in Geographic Information Science,
https://doi.org/10.1007/978-3-030-06058-9_3

Seymour et al. 2004) indicating the compelling stimulus of such courses for choosing a career in science or academia, pursuing graduate schools or simply improving the personal development or growth of self-confidence in students (Taraban and Blanton 2008).

Geographic information science (GIS) and remote sensing (RS) courses situate well with this teaching challenge to introduce and expose students to the scientific research process. Typically, GIS and RS courses have theoretical and practical components where a computer lab is a necessary learning component and provide students with an exposure to sophisticated software and sometimes computer programming (Arribas-Bel and Reades 2018; Bowlick et al. 2017). In upper-division courses, there are usually group-based or individual projects as part of the learning experience where students must assemble geospatial data, conduct relevant analysis, and communicate the results (Delahunty et al. 2012; Kemp et al. 1992).

The main objective of this study was to investigate how undergraduate students enrolled in an upper level project-based spatial modeling course can perform and be involved as researchers and scientists. The driving motivation was to encourage students to think and express themselves as scientists by being engaged at real conference settings to communicate the findings for their research projects (e.g., simulation model results) and enrich their experience in scientific communication and interactions with the conference participants as colleagues and academics. Anonymous questionnaire feedback was used to assess student perceptions about the learning experience and value of the research skills obtained.

Background

The Spatial Modeling Course

In the Geography Department at Simon Fraser University (SFU), spatial modeling (GEOG 451) is an upper-division course offered within the spatial information science (SIS) stream and focuses on the theory of complex systems, geographic information systems (GIS), and their use for building geosimulation modeling approaches to represent dynamic geographic phenomena using raster-based GIS software. The course is strategically positioned between the seminar-type and knowledge-type formats to give students optimal learning benefits. In the course, students are required to conduct literature reviews that are rigorous in order to be prepared for class discussions and knowledge exchange. However, depending on the student enrolment, this strategy is not always effective for student learning. When the class size is too large, not all students are sufficiently engaged in discussion exchanges. When the class size is too small, traditional lecturing may not be engaging enough for some students. Moreover, the complexity of the subject matter makes the materials challenging for students to learn by only reading scientific papers as in a regular seminar course, which in this case may be too advanced for students who are just

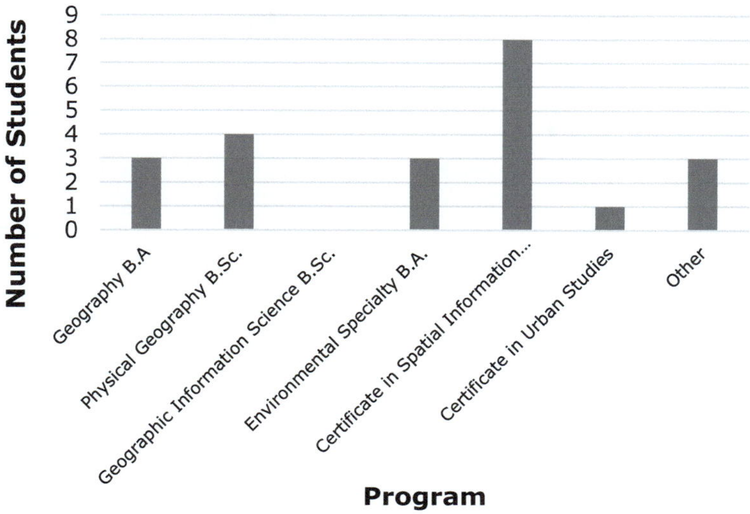

Fig. 1 Distribution of the program of study students are enrolled into

beginning to learn the concepts. One solution is students need to obtain both knowledge transfer through the traditional classroom lecture and practical experience through computer laboratory delivery (Delahunty et al. 2012). This creates a dual character for the course: one that is about learning and acquiring the knowledge on new topics and the other about learning that is practically oriented and structured by the scientific research process (Grant et al. 2013; McMillan et al. 2018).

The study was implemented in the September–December 2015 running of the GEOG 451 course. There were 14 students enrolled in the course with 9 males and 5 female students. The students were from a variety of academic backgrounds (Fig. 1). The majority of students took the course to complete their certificate in spatial information systems (SIS) and also were enrolled in other degrees. There were some students in other programs not listed in the introduction survey including the student exchange program, a BSc double major including arts and technology and geography. Most students took the course out of interest, although some took the course to meet their program requirements.

Embedding the Scientific Research Process

During the spatial modeling course, the knowledge content was specifically structured to follow the stages of the scientific research process: (1) literature search, (2) collecting and managing geospatial data, (3) building geosimulation models, generating and analysis of model output results, and (4) communication of results. Table 1 shows the details of the course structure with the scientific research process

Table 1 Course structure and the scientific research process

Scientific research process	Knowledge-style components	Seminar-style components
Literature search		Reading academic papers
Collecting and managing geospatial data	Instructional computer labs	Traditional lectures
Building and implementation of the geosimulation model, generating and analysis of model output results	Instructional computer labs	Traditional lectures
Communication of results		Poster and oral presentations, scientific paper

Table 2 Specific research questions investigated

Question(s)	Data or information source
1. What are the students' learning outcomes on the overall knowledge acquired from the course content?	Pre-post test
2. How students perceive the scientific process and what they have learnt about it?	Opinion survey 1
3. How students' experience from scientific communication (poster, oral presentation or written report) and conference attendance have reinforced their ability to think and act as scientist?	Opinion survey 1
4. How the experiences on learning about research process and taking this particular course have overall influenced the students to pursue graduate studies?	Opinion survey 2

embedded. The traditional lectures were on the theoretical concepts of complex systems and modeling approaches such as cellular automata and agent-based modeling. The instructional computer labs were on the practical use of GIS raster-based software Idrisi TerrSet (Clark Labs 2015) and ArcGIS (ESRI USA 2015) with hands-on experience on how to implement the geosimulation model with the software. The communication of results was threefold: (1) present posters in a real conference setting, (2) deliver oral presentations in a hypothetical conference setting at the classroom venue, and (3) prepare scientific paper as a written report of the developed modeling approach and obtained results.

Specific Research Questions

The specific research questions and the corresponding evaluation instruments for this study are outlined in Table 2.

Results and Discussion

Learning Outcomes on Overall Knowledge Acquired

Students were given identical tests once in the beginning of the semester and once near the end. The test was composed of knowledge-based questions covering a variety of materials discussed in the course lectures and the lab. Each question also required that students record their confidence in their answer and where they believed they obtained this information. Pre-post test responses from 12 students were analyzed from a class size of 14 students. Two students who took the pre-test did not take the post-test and thus were excluded from the analysis.

Students were informed that the pre-post test had no impact on their course grades. Consequently, the result is a better indication of actual knowledge retention instead of retention for the purposes of writing a test. On average, students performed better on the post-test. Overall, the average for the pre-test was 29%, and the average for the post-test was 41.6%. Figure 2 compares the grade achieved on the pre-test with the grade achieved on the post-test for each of the students. Each student performed better, some with significantly more improved grades than others, and two student's pre- and post-tests had the same grade. The overall increase in grade may be attributed to the fact that most of the material in the pre-post test was not taught until after the pre-test was distributed. Students indicated that the majority of their information was acquired in the course lecture.

In general, students obtained improved marks on questions addressing major theoretical course themes, often highlighted during the lectures, for example, questions Q1, Q3, and Q4 (Fig. 3). Students also correctly answered questions that put the theory learned in the lectures into practice in the lab for questions Q6 and Q7 on model calibration, validation, and map comparison methods (Fig. 3). Question

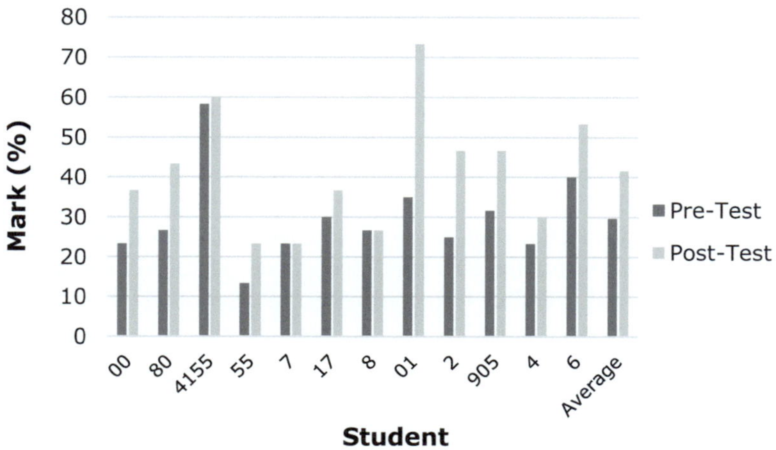

Fig. 2 Comparing the average mark between the pre-test and post-test for each student

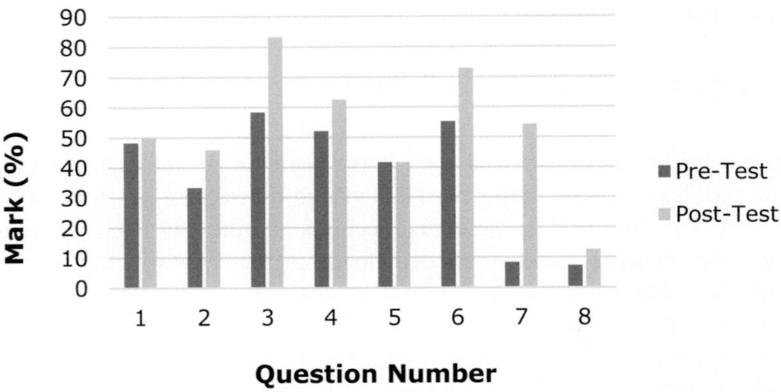

Fig. 3 Comparing the average mark per question for the pre-test and post-test

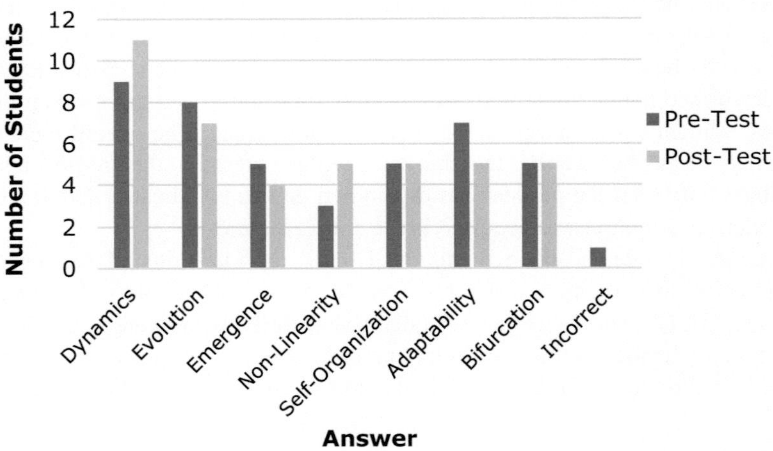

Fig. 4 Analysis of the retention of concepts over time

Q8 (Fig. 3), consistently answered incorrectly, was related to the mathematical formulation of cellular automaton, which may reflect typical geography students' apprehension and perceived challenge of mathematics as identified in the survey.

Understanding the characteristics of complex systems (Q1) was an important theme discussed in the lectures before the pre-test and for the duration of the course. Students were asked to list the characteristics. It was useful to determine which concepts were retained over time from the pre-test to the post-test (Fig. 4). Overall, students understand that complex systems are dynamic and evolve, but they tended to forget about the other characteristics such as emergence, nonlinearity, and adaptability.

The change in student confidence over the term was also explored using coefficient of determination (R^2). The R^2 value provides a measure of the extent to which the dependent variable is explained by the independent variable(s). During the

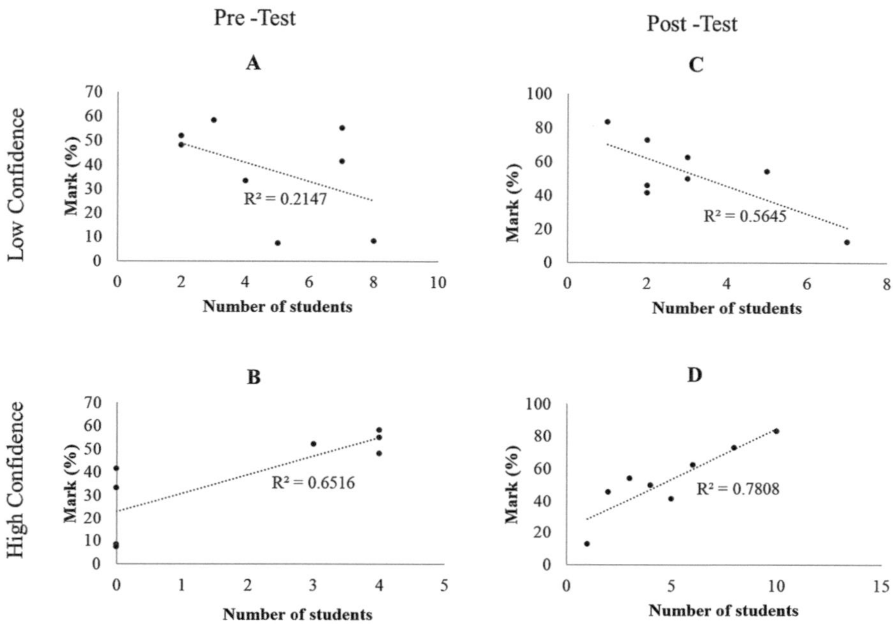

Fig. 5 Correlation values between marks in percentage (%) and number of students reporting low and high confidence levels in the pre- and post-tests

pre-test, low grades were negatively correlated with the number of students declaring low confidence (Fig. 5a), although the association was weak ($R^2 = 0.215$). This means that naturally, as grades decrease, the number of students with low confidence increases. The weak correlation is associated with some cases where as grades decrease, the number of students with low confidence decreases indicating overconfidence in the beginning of the term. In addition, the results indicate there is a positive correlation with high grades and the number of students declaring high confidence (Fig. 5b), with strong association ($R^2 = 0.652$). This means that naturally, as grades increase, the number of students with high confidence increases. There are some instances where there are no students with high confidence associated with relatively high grades, indicating under confidence. It may be concluded that students in the beginning of the term don't have a clear grasp of their own knowledge level.

During the post-test, the negative correlation between low grades and the number of people with low confidence became significantly stronger where $R^2 = 0.565$ (Fig. 5c). In addition, the positive correlation between high grades and high confidence became significantly stronger where $R^2 = 0.781$ (Fig. 5d). This gives a strong indication that students became more aware of their knowledge level regarding course content. Under-confidence and overconfidence were less prevalent. In addition, there is an inverse trend in the confidence level of the students between the pre-test and the post-test. In the pre-test, low confidence dominated, whereas in the post-test, high confidence dominated, with medium confidence falling in the middle

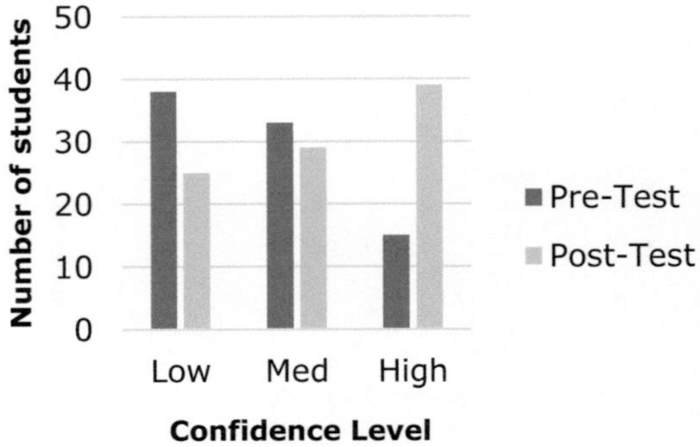

Fig. 6 Comparing the change in perceived confidence levels between the pre-test and the post-test

(Fig. 6). During the progress of the course during the term, students improved on their knowledge of the course materials, and their confidence level has increased.

Perceptions About the Scientific Process and Lessons Learned

During the term, as more concepts and lectures were introduced, students were given the opportunity to experience the major steps in the process of scientific thinking and research, so they could complete their research projects. These steps include identifying the spatial dynamic phenomena to study, defining the problem and research questions, conducting a literature review, developing a model, implementing the model with data and with use of raster-based GIS software, analyzing the results of the model, and communicating the results of a model through poster, oral presentation, and scientific paper writing. A special lecture has been added to particularly inform students about the scientific research process. As presented on Fig. 7, overall results indicate that students reported that they either had an excellent, very good, and good understanding of each of these steps at the end of the course. The least understood steps include conducting a literature review, developing a model, and implementing the model with data, each of which had one student with a poor understanding. The most success was felt in analyzing the results of the model and communicating the results, although these steps also had a few students who expressed only fair understanding of these concepts.

Based on the comments provided in the online Survey 1 about what students were most looking forward to learn in the course, the steps about developing a geosimulation model were more preferred than the ones related to literature review, poster presentation, or the individual work on the project (Fig. 8). The students indicated that they preferred the work on developing and implementing the actual model, as this was something that they could clearly see accomplished by the end of the course and that they enjoyed being in that process throughout the course. However, students were challenged the most with the problem formulation, model creation, and finding adequate geospatial data for their modeling project (Fig. 9), all of which are very real challenges experienced in postgraduate academic research.

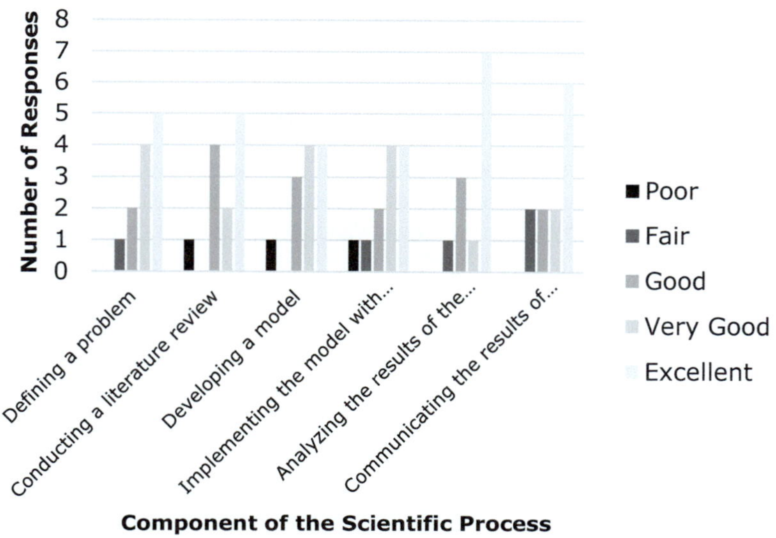

Fig. 7 Levels of understanding of the scientific process

Fig. 8 Components in the course work that students enjoyed

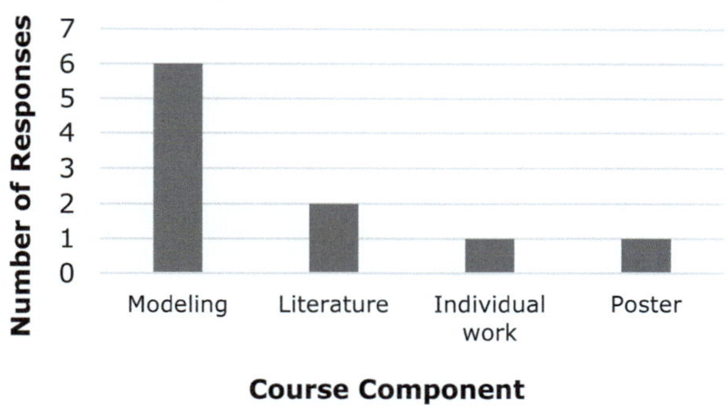

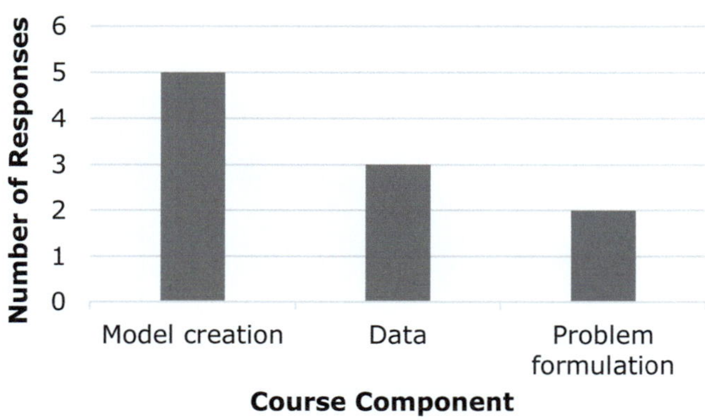

Fig. 9 Components in the course work that students perceived as challenging

Experience from Scientific Communication and Conference Attendance

One of the factors that may have contributed to the success in communicating the results is the variety of mediums through which the students were given an opportunity to use. Students put together a poster presenting their research for a display at the 2015 ESRI User Conference, held in Vancouver, Canada (Fig. 10). Students spent the day at the conference listening to various sessions from talks introducing novel technological development related to the GIS software and achievements obtained from GIS practitioners from industry and government to those related to the GIS-related research sessions with scientific findings. The conference not only gave them a chance to present their own research, but also to experience other researcher's presentations, to engage in discussion of scientific findings and to network with researchers and practitioners in the field. In addition, the conference improved their knowledge of GIS and spatial modeling and provided a better understanding of the variety of jobs available. Comments provided by the students in the Survey 2 (Fig. 11) highlight their enjoyment of presentations and of learning about the latest GIS technology. Some felt that the conference was too application focused/ESRI focused or that the length of the conference was too long. Fortunately, four students had the opportunity to make a new connection at the conference. This exposure of the students to professionals and academics who work in the industry is important for potential future job search by students.

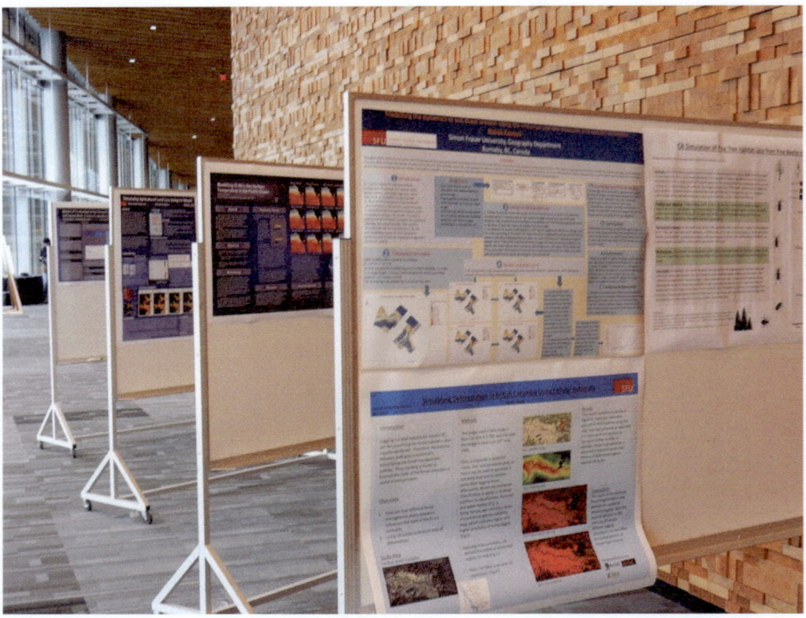

Fig. 10 Student posters at the conference event

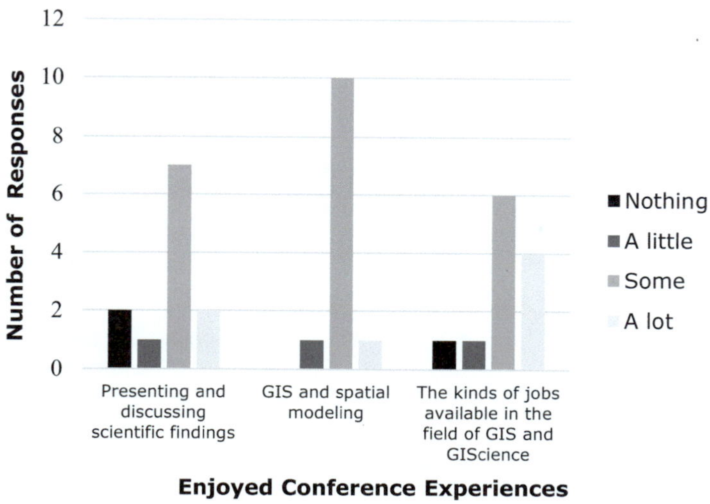

Fig. 11 Conference experience

Learning About the Research Process and Influence on Graduate Studies

The students were asked about their future plans including the following: (1) pursuing graduate studies in GIS, (2) pursuing graduate studies in another field, (3) pursuing an academic or scholarly career, or (4) pursuing a career in GIS outside of academia. Based on the surveys conducted at the beginning, near, and after the course was completed, the changes of their plans over time were presented for the pre- (Fig. 12), near-post- (Fig. 13), and post- (Fig. 14) spatial modeling course offering.

The results indicate that for the pre-course offerings (Fig. 12), students were most likely going to pursue a career in GIS or GIScience outside of academia; however, there was a larger group of students that were also interested in pursuing graduate studies in GIS. It is also worthwhile to note that there were some students who found it unlikely that they would pursue graduate studies in another field. This indicates that students were most likely taking this course as aid in their future GIS careers regardless of whether it was in academia or industry.

The near-post-course offerings results (Fig. 12) indicate a slight shift from students who were likely to pursue a career in GIS of GIScience outside of academia to somewhat likely. Results also depict a shift in pursuing graduate studies in GIScience to from unlikely to somewhat likely. What can be seen here is that less students are interested in working in industry and more interested in academia near the completion of the course. Finally, as presented in Fig. 13, the post-course analysis indicates an increase in interest in pursuing a career in GIS and GIScience outside of academia from somewhat likely to likely. There is also an increase of students to likely pursue graduate studies in GIS or GIScience. The distribution is more

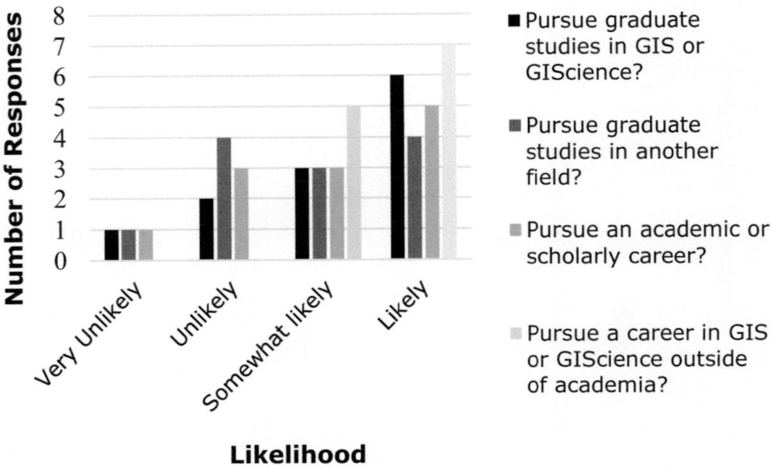

Fig. 12 Future plans pre-course

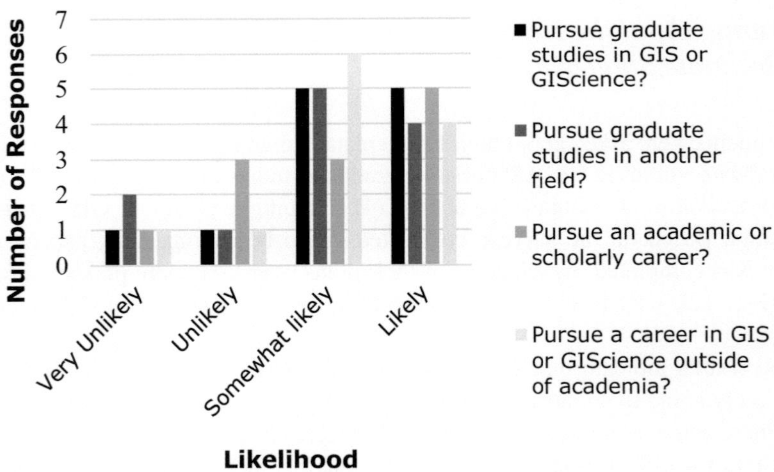

Fig. 13 Future plans near-post-course

GIScience, and contributes some of their decision to what they learned in the spatial modeling course and their experience taking courses in the geography department; (2) student 2 has two semesters remaining, applied to a graduate program in GIS and GIScience, and taking this course or other courses in the geography programs did not contribute to their decision; and (3) student 3 has 1 year or more to go and has plans to apply for graduate studies both inside and outside of the field of GIS and GIScience, and this decision was related to the experience taking spatial modeling course and other courses offered within the geography department program.

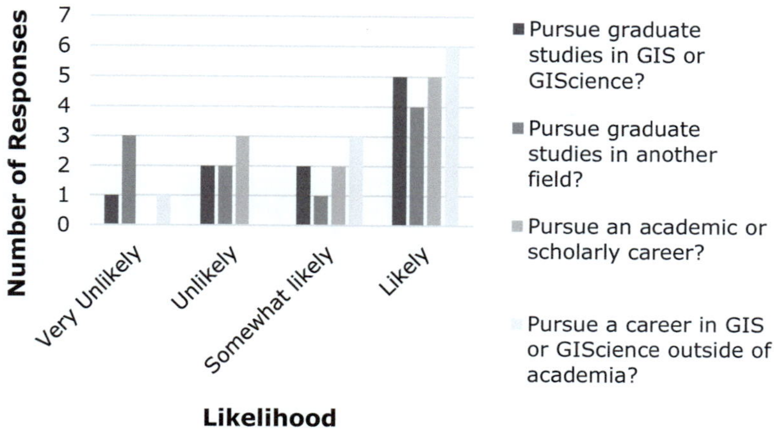

Fig. 14 Future plans post-course

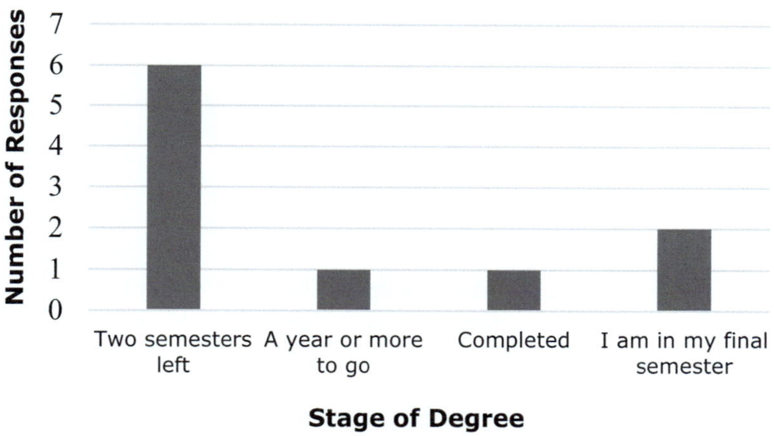

Fig. 15 The length of time before undergraduate program completion

similar to the pre-course results but of interest is that the number of people to pursue graduate studies in another field increased to very unlikely. This can be attributed to the overall conference experience and interactions with practitioners in the field. Students would like to engage in future work in the field of GIScience and GIS; however, the tension and flip-flopping between staying in academia or getting an industry job is highlighted over time.

The majority of students who took the spatial modeling course (GEOG 451) have two semesters left before graduation. However, two are in their final semester, and one has completed their degree (Fig. 15). Of the ten students who completed Survey 2, three have applied or have plans to apply for graduate school: (1) student 1 is in their final semester, applied to a different program outside of GIS and

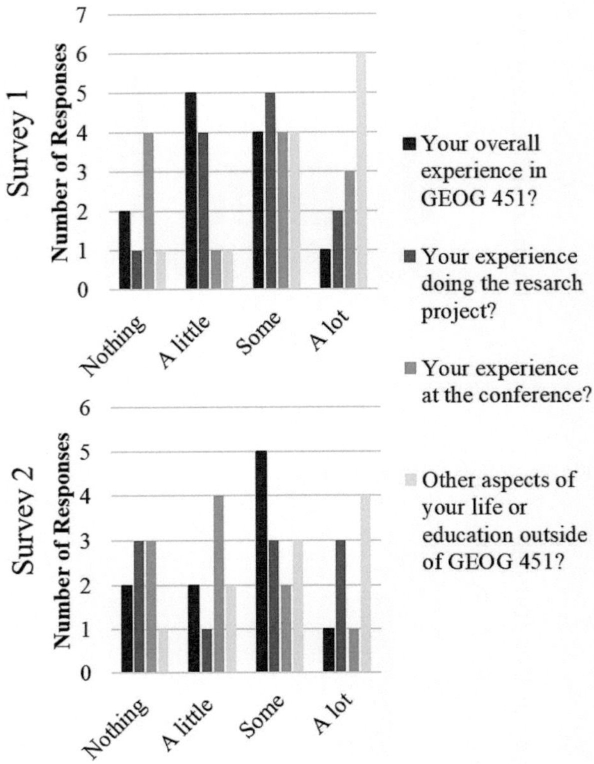

Fig. 16 Results of Surveys 1 and 2 presenting the different factors influencing students' future plans

Overall, the experience in spatial modeling (GEOG 451) course had a little to some influence on students' future plans in Survey 1, although this influence increases to some in Survey 2 (Fig. 16). This indicates that students had time to reflect on their learning experience in the course and their experience was influencing their decisions regarding their future. It appears that students mostly attribute other aspects of their personal life to contributing to their decisions regarding their future on pursuing graduate studies with research-oriented academic career or finding a job in the industry. Very few students contributed their decisions regarding their future plans to the conference experience. However, some students indicated that their conference experience contributed to their decision a lot and also indicated that they plan to pursue a career in industry, which suggests that a positive conference experience created excitement about working in GIS industry.

Conclusion

The dominant result of the study indicates that the students' experience, including learning about the scientific research process, the ability to communicate research results, and participation at a conference settings, had some impact to reinforce their ability to think and act as a scientist. The results from student responses suggested that more interaction with a scientific audience may lead to an enhanced development of scientific research skills. Given the relatively small number of students enrolled in this course, a future study will be necessary to explore the consistency of the previous results using comparisons across two or more course offerings. Moreover, some of the positive aspects of this research study are the overall enhancement of the course offering by introducing additional topics on scientific research process and the delivery of project posters as additional way for students' communication of the modeling project results. Overall the experience gained by students was rewarding, and the research skills gained in GIS, GIScience, and geosimulation will be a lifelong asset.

Acknowledgments This project was fully funded by a Simon Fraser University (SFU) Teaching and Learning Development Grant (TLDG) awarded to the first author. The authors thank Angela McLean and Laura D'Amico from the Institute for the Study of Teaching and Learning in the Disciplines (ISTLD), SFU, for their valuable guidance during the project development and implementation. Special thanks to Dr. Brent Hall of Esri Canada for making possible the students' attendance and poster presentations at the 2015 Esri User conference in Vancouver, Canada.

References

Arribas-Bel, D., & Reades, J. (2018). Geography and computers: Past, present, and future. *Geography Compass, 12*(10), e12403.

Bowlick, F. J., Goldberg, D. W., & Bednarz, S. W. (2017). Computer science and programming courses in Geography Departments in the United States. *The Professional Geographer, 69*(1), 138–150.

Chen, X. M. (1998). Integrating GIS education with training: A project-oriented approach. *Journal of Geography, 97*(6), 261–268.

Clark Labs. (2015). *TerrSet geospatial monitoring and modeling software*. Worcester, MA: https://clarklabs.org/terrset/

Delahunty, T., Lewis-Gonzales, S., Phelps, J., Sawicki, B., Roberts, C., & Carpenter, P. (2012). Land cover change and remote sensing in the classroom: An exercise to study urban growth. *Journal of Geography, 111*(3), 113–122.

Ebenezer, J., Kaya, O. N., & Ebenezer, D. L. (2010). Engaging students in environmental research projects: Perceptions of fluency with innovative technologies and levels of scientific inquiry abilities. *Journal of Research in Science Teaching, 48*(1), 94–116.

ESRI USA. (2015). ArcGIS Desktop, release 10.4. Redlands, CA: http://desktop.arcgis.com/en/

Grant, M. C., Fisher, D., & Lapp, D. (2013). *Teaching students to think like scientists: Strategies aligned with common core and next generation science standards*. Bloomington: Solution Tree.

Kemp, K. K., Goodchild, M. F., & Dodson, R. F. (1992). Teaching GIS in geography. *The Professional Geographer, 44*(2), 181–191.

Lopatto, D. (2007). Undergraduate research experiences support science career decisions and active learning. *CBE Life Sciences Education, 6*(4), 297–306.

McMillan, C., Loads, D., & McQueen, H. A. (2018). From students to scientists: The impact of interactive engagement in lectures. *New Directions in the Teaching of Physical Sciences, 13*(1). https://doi.org/10.29311/ndtps.v0i13.2425.

Seymour, E., Hunter, A.-B., Laursen, S. L., & DeAntoni, T. (2004). Establishing the benefits of research experiences for undergraduates in the sciences: First findings from a three-year study. *Science Education, 88*(4), 493–534.

Taraban, R., & Blanton, R. L. (Eds.). (2008). *Creating effective undergraduate research programs in science: Transformation from student to scientist*. New York: Teachers College Press.

Trosset, C., Lopatto, D., & Elgin, S. (2008). Implementation and assessment of course-embedded undergraduate experience: Some explorations. In R. Taraban & R. L. Blanton (Eds.), *Creating effective undergraduate research programs in science: The transformation from student to scientist* (p. 256). New York: Teachers College Press.

The AMETHYST Program: The NSERC CREATE Experience

Craig Coburn

Abstract This paper reports on the experience of the AMETHYST program at the University of Lethbridge. The Advanced Methods, Education and Training in Hyperspectral Science and Technology program was awarded an NSERC CREATE in 2010 as one of the first of these grants awarded to Canadian universities to focus on the training of students in science. The purpose of the NSERC CREATE program was to support training students (at various levels from undergraduate to post-doctoral) to expand opportunities by funding nontraditional training programs. The AMETHYST program has trained over 30 students and created a program that provides an enriched experience in hyperspectral imaging science and remote sensing. The program included both student professional development and collaborative experiences beyond what is normally experienced in cooperative education models. This model was developed with the view to program continuation past the grant's end point. While the program was successful, NSERC's attitude towards funding graduate students fails to meet the future needs for training HQP in Canada.

Keywords Geography education · Remote sensing · GIS teaching · Research funding · Cooperative education

Introduction

One of the most pressing challenges that face the post-secondary education landscape is the diminishing financial commitment of federal and provincial government funding sources towards university research. These reductions in funding levels have been felt by all post-secondary levels but are especially acute in the science-related disciplines, where reliance on graduate students as integral parts of the machinery of science is essential to ensure continued research productivity.

C. Coburn (✉)
Department of Geography, University of Lethbridge, Lethbridge, Canada
e-mail: craig.coburn@uleth.ca

S. Balram, J. Boxall (eds.), *GIScience Teaching and Learning Perspectives*, Advances in Geographic Information Science,
https://doi.org/10.1007/978-3-030-06058-9_4

The Canadian federal government funds science and engineering research through grants provided by the Natural Sciences and Engineering Research Council of Canada (NSERC). The NSERC has a wide variety of funding mechanism to provide funding to students (at the undergraduate through post-graduate levels) as well as researchers (Discovery Grants, Discovery Accelerator Supplements) and to groups of researchers (CREATE Grants). These funding mechanisms are designed to provide support in broad area of science and engineering with the stated objective of making Canada a world leader in post-secondary education research and to create a prosperous knowledge economy through the training of a highly skilled workforce (NSERC 2013).

For the past 50 years, the Canadian federal government has made significant investments to the post-secondary education sector by increasing funding through various funding agencies as well as providing essential infrastructure through an increase in the number of universities and colleges. The overall objective has been to drive Canadian economic growth by providing access to post-secondary educational opportunities. The additional societal benefits of increasing social and economic equality recognize the philosophy that access to post-secondary education opportunities in Canada is based on student ability rather than economic means.

Reductions in the financial commitment coming to universities from the federal and provincial governments have led to significant reductions in the operating resources of most post-secondary institutions with students now contributing over 23% of the total revenue for universities in 2012 (latest source information from Statistics Canada) as a national average. An average increase in tuition of 26% over the 2007–2013 period for undergraduates is a direct result of the reductions in funding from governments. These changes to the funding model create a number of complications as cuts to funding are coincident with increasing technological, societal and demographic changes with increased emphasis on the post-secondary system to develop highly skilled personnel (Canadian Federation of Students 2013).

The increasing demands of the labour market university graduates with enhanced skills have broken down geographic boundaries and created a market where skilled people from many countries are competing for limited numbers of positions. These positions require more qualifications and often require vetting by national or provincial professional bodies charged with vetting qualifications. These factors are increasing the pressures on the post-secondary sector to develop programs that are focused on employment marketability and future-proof students.

In Canada, the NSERC Discovery Grant (DG) has been the cornerstone for the funding of science at universities as there is no built-in research funding. The objective of the DG is to provide a base of funding for research programs and to provide flexibility for researchers to pursue areas of research where directed deliverables may not provide the most rapid advance in science. It is also meant to fund graduate student training to provide the next generation of science leaders.

Recent changes to the way that funding is allocated in the DG have resulted in a decrease in overall number of grants awarded with success rates falling to between 60% and 70% of applicants. This reduction in base research funding has increased the pressure on both the applicants and the funding agency. The latest thrust for

NSERC is for the funding agency to fund fewer programs at an adequate level but not provide sufficient funds to fully support academic research. This is a remarkable divergence for the main provider of science funding in Canada and shifts the focus of the funding envelope away from traditional sources to other funding agencies. It has been noted that this approach to funding favours larger institutions, limits undergraduate participation in research, adversely impacts smaller universities with heavier teaching loads and also adversely impacts young scientists who are attempting to start research careers. The emphasis on "stars" no longer provides for the incubation of academic careers – "stars" are selected at very early career phases (Joos 2012).

Given this change in the funding landscape from NSERC, the introduction of a new funding mechanism that focused on meeting the government's objectives and provided long-term funding for scientists was a very attractive opportunity. The stated goals of the Collaborative Research and Training Experience Program (CREATE) was to improve the mentoring and training environment for graduate students by improving efforts in areas such as professional skills development (primarily communication and collaboration) and for the students to gain valuable experience relevant to both academic and nonacademic research environments. The program began in 2009 and as of 2015 has awarded 122 grants for an average of $1.65 million dollars each over 6 years for a total funding envelope of $201.3 million dollars. In 2015, the NSERC budget for Discovery Grants was $340 million dollars, and the CREATE program granted $28 million dollars for 2015 or around 8.25% of the DG allocation total to only 17 research groups nationwide.

The NSERC CREATE foundations are to provide additional mentoring and training for students by moving them between industry/government and academic settings. The programs that have been funded must be interdisciplinary in nature and were meant to increase the collaboration between industry and academia with the eventual goal to improve "job readiness" of graduating students.

While NSERC offers no official theoretical framework within which the CREATE program resides, there is little doubt that experiential learning (Dewey 1938) is the governing framework. Dewey (1938) asserted that while all learning is based on experience, not all experience is educational. The overall emphasis of this model is "learning by doing" with specific focus on field trips, problem-solving, supervisor-mentor instruction methods and other hands-on learning dominating the method of knowledge transfer where learners construct meaning from their experiences (Doolittle and Camp 1999).

The NSERC AMETHYST CREATE program was situated in the experiential learning structure proposed by Kolb (1984) where there are cycles of learning with four stages. Within this model, learners are ever cycling between (1) reflective observation and (2) active experimentation and (3) concrete experience and (4) abstract conceptualization. These cycles can take many forms. For example, active experimentation leads to concrete experience, and that leads to reflective observation and then abstract conceptualization. Though the learning pathways are not always linear, and the linkages not always direct, knowledge is accrued over time. From this theoretical framework, the educational objectives and methods used were developed for this program.

AMETHYST CREATE Program

In 2010, the researchers at the University of Lethbridge decided to apply for a CREATE grant to assist in our mission to provide the best possible education, research and funding opportunities to our students in the field of remote sensing. The proposal was successful, and we launched the Advanced Methods, Education and Training in Hyperspectral Science and Technology (AMETHYST) program in March 2010. The University of Lethbridge had made several strategic hires that resulted in a reasonably large number of remote sensing scientists (five core faculty members in 2010) and had established Canada's only undergraduate degree in remote sensing.

With a range of research interests spanning the science of remote sensing, including a corporate entity (the Alberta Terrestrial Imaging Center – ATIC), the University of Lethbridge was well placed to succeed with our grant application. Added to the benefit list were several local businesses focused on geospatial technologies as well as a government research facility (Agriculture and Agrifood Canada, Lethbridge Research Centre). Our application was also enhanced by the inclusion of medical imaging as a complementary area to the core remote sensing focus as part of a broader vision for imaging science and technology at the university.

In the overall global context, imaging science and technology is a major tool for a wide variety of scientific endeavours. Our program proposal focused not only on the unusually high level of expertise in the faculty at the university but on the growing job market and Canada's role in the global imaging landscape. At the time, the Canadian government had an interest in hyperspectral imaging for terrestrial image analysis, and there was a need for highly qualified people to fill this potentially expanding area of imaging science.

The main objectives of AMETHYST were to:

- Provide unique theoretical and experimental training for students and new researchers. The AMETHYST experience will provide graduates with the right mix of broad theoretical and experimental training and solid training in professional skills that will enable them to pursue science and technology careers in industrial, academic or government settings.
- Establish a pathways program of local, national and international internships, workshops, events and workplace assignments, tailored to each trainee, that provide leading-edge yet hands-on and individualized training in interdisciplinary research and that serve to create contacts between trainees and the workforce.
- Build on existing strengths including the new multidisciplinary major in remote sensing, the only remote sensing degree program in Canada. Extend this program to include graduate degrees in remote sensing.
- Provide a Canadian solution to Canada's Department of National Defence (DND) requirements for a two-year M.Sc. in imaging science (not available in Canada; DND trains individuals in the USA).

AMETHYST Structure

The AMETHYST program is based on five key components: (1) multiple education levels, (2) blended learning environments, (3) professional skill development, (4) interdisciplinary partners and (5) defined programming outcomes (Fig. 1). The program was designed to prepare trainees for the highly interdisciplinary nature of modern imaging science and technology and provided trainees with tools that the NSERC CREATE program focused on by providing them with collaborative and integrative educational experiences in interdisciplinary multi-sector contexts.

We have called this approach the *AMETHYST Pathways Program* which was based on an experiential learning model, where students were given opportunities to explore a range of available options and learn directly from experts in a variety of settings (academic, government and industrial). This exposure was a key component of NSERC's CREATE program as they were willing to fund programs that ensured motivated and highly qualified persons, with workforce-ready skills.

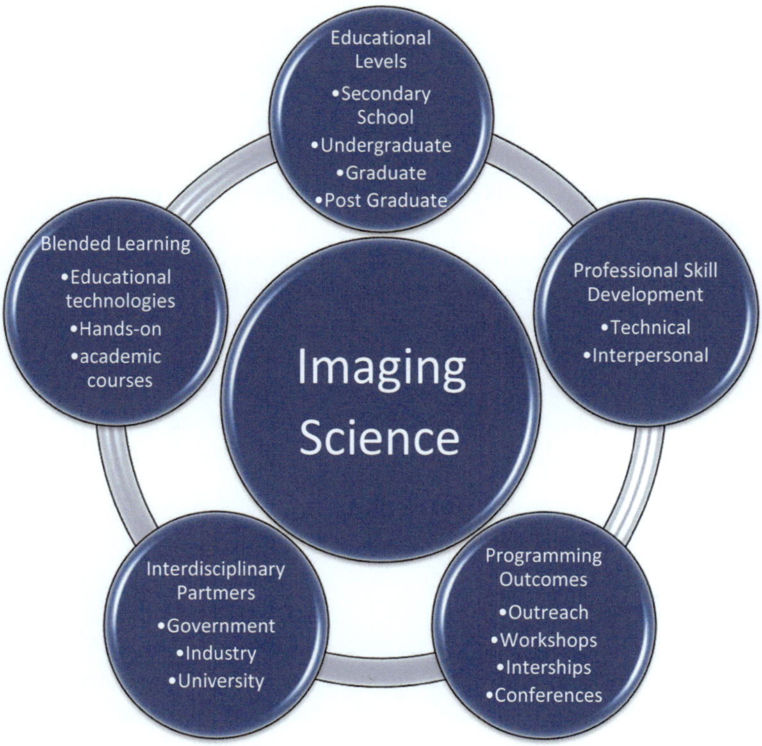

Fig. 1 AMETHYST program structure

The AMETHYST Pathways Program

Secondary School Outreach

The concept of the pathways program was to initiate contact with secondary school students across southern Alberta and encourage them to come to the University of Lethbridge to pursue an education in remote sensing. This was accomplished by contacting guidance counsellors in the local school system and having classes tour our lab facilities. There was an existing relationship between physics teachers and our physics and astronomy department, and we entered into an agreement that our labs/demonstrations would be part of their outreach program. We also occasionally took on secondary students to work in our labs to provide them with some exposure prior to beginning their university careers.

Undergraduate Pathways: Internships

The AMETHYST Pathways Program was designed to draw upon our history of extensive involvement with government and industry organisations and other academic institutions conducting research or developing applications in imaging S&T. A formal program was established that provided cooperative education style placements with industry/government and research laboratories across Canada and around the world. Undergraduates admitted into the program were guaranteed three summer placements (one placement for each summer of their undergraduate education). This guarantee was attractive to students as they didn't have to worry about finding summer employment for their entire degree program.

The program was designed so that each student would spend their first placement at the University of Lethbridge gaining a base level of competence in the field and then spend the subsequent two summer placements in either industry or government settings. This unique exposure to a broad range of possibilities provided exposure to a range of employment opportunities. It was also hoped that this level of commitment to our collaborators would encourage them to continue to engage with our research group past the termination of the NSERC CREATE program.

Annual Workshops

The AMETHYST Pathways Program included two separate types of workshop to assist trainees with research training objectives and the clarification of their specific career goals. The workshops were based on a philosophy of "blended learning", whereby students were exposed to diverse learning modalities, environments and technologies that enhance and accelerate the learning process. Both workshop types took advantage of new instructional technologies such as on-line interactions and dynamic feedback from workshop participants.

- *AMETHYST Workshop on Hyperspectral Imaging Science and Technology*: This annual workshop technology is a 2-day long intensive interdisciplinary research training experience open to students and researchers across Canada. The workshop theme changed from year to year, and experts in the field were brought in from our AMETHYST partner institutions and other invited prominent researchers and experts.
- *AMETHYST Workshop on Career Development and Workforce Preparation*: This annual workshop offered trainees a unique and diverse training experience by allowing them to acquire a full scope of individual and team-oriented professional skills that are highly valued in, and optimized for, the job market. This workshop focused on presentation skill development, project management, time management in work and research environments and written communication skills.

Graduate Pathways: Internship in Hyperspectral Science and Technology

AMETHYST post-graduate trainees will broaden their horizons in one or more partner laboratories, with internship opportunities in both computational studies and in the use of the specialized instruments and techniques. When the regulations of the institutions involved permit, our trainees will be offered the possibility of obtaining an international, jointly supervised graduate degree granted as a result of studies at two or more participating institutions.

Graduate Pathways: Short-Term Internships

Graduate studies are normally conducted on the individual student/professor mentorship model. With the AMETHYST program, we introduced the concept of cooperative training with the goal of providing the student with a training experience directly related to their graduate research. This internship was hosted by government, industry or university laboratories and provided additional networking and important skill development for our students. Students at the master's level would be awarded a single internship, while Ph.D. students would receive two research internships. As many of the students at the graduate level would be joining the AMETHYST program for the first time, the skills and technical workshops would also be available as well as the formulation of a professional development strategy and plan for each student.

AMETHYST Outcomes: Evaluating the Program

The management of the program was an ongoing annual event, so there were ample opportunities to assess various key performance indicators (KPIs) like number of students in the program, successful internship placements, conference presentations and refereed publications. In general, the program was able to meet these main KPIs, and while there were struggles to get the number of students in the program to meet the initial estimates, these estimates were based on our best educated guesses, rather than long-term success running a unique training program of this type. Table 1 presents the estimated and actual number of students that participated in the program from 2010 to 2016 (program termination).

There were numerous challenges to offering this program at the University of Lethbridge. The underlying assumption of the NSERC CREATE initiative is that the post-secondary system of training does not provide workforce-ready graduates and that by combining work experience with traditional academic programs, a better prepared student would emerge and be ready to become a productive worker upon graduation.

This approach worked with the same degree of efficacy as the current cooperative education model that many post-secondary institutions already use for undergraduate placement – which is to say that it had the intended effect as the students were able to gain valuable work experience and having the funding from NSERC ensured diversity in placement as the employer didn't have to contribute to the funding if they were unable or unwilling to do so.

The graduate experience was suboptimal. The concept that was discussed with partners in industry and government was to have reasonably short internships where the students would be acquiring specific job skills; unfortunately this is not what actually happened as employers wanted the students for similar lengths of time as undergraduate interns who were employed for at least one semester. This length of time was needed as graduate students were not working on their individual thesis-based research but on unrelated tasks.

Industry failed to recognize the value of the experiential learning and blended training model. In general, the industry partners saw graduate internships as "free-labour" and wanted students to complete meaningful projects that were in addition to their thesis. This resulted in longer than anticipated times to completion for students in the AMETHYST program. Industry would also freely cancel commitments to accept students often at the last moment and without warning.

The longer time to completion and difficulty with finding willing industry partners was a consistent problem throughout the program. In the end, there was

Table 1 Number of student trainees

Trainee intake	Estimated	Actual	Difference
Secondary	0	3	3
BSc	22	10	−12
MSc	17	24	7
PhD	7	10	3
PDF	6	4	−2

little difference in the success of graduate students in the job market. Our graduate students already experience very high success rates in relevant placements within all sectors. There was no indication that the students that had completed this program were appreciably different in their job readiness or other skills that NSERC was targeting.

Conclusion

The AMETHYST program has trained over 30 students and created a program that provides an enriched experience in hyperspectral imaging science and remote sensing. One of the goals of the NSERC CREATE programs was to engage industry in the training process so that the students that they hire would be more "job-ready". To this end, part of all NSERC CREATE programs is to ensure that the program continues without continued NSERC support. Perhaps the biggest failure on the part of the funding agency was the failure to recognize that industry wasn't aware of these programs and they have, in our experience, little interest in continuing to fund what is primarily and educational program. Therefore, all of the effort that went into the creation of this program will most likely fail to produce a long-term stable funding source.

The program included both student professional development and collaborative experiences beyond what is normally experienced in cooperative education models, and in total, we have had more graduate participation in this program than we had had total over the proceeding decade in our department. This does indicate that there is willingness of students to seek out this higher level of education, providing it's appropriately funded. This runs contra to the current NSERC stance that minimal funding to support university research will yield funding from other sources and that we will be able to meet market demands for HQP.

The experiential learning experience provided by this program tried to emphasize the learners' active role in the creation of knowledge through experience, reflection, concrete examples and theoretical abstractions. Within this framework, the degree of engagement of the learner is central to the process of experiential learning, and as with most educational endeavours, there are degrees of success with engaged students gaining required skills and flourishing and those that are unfamiliar with this level of commitment often languishing and some exiting the program altogether.

There is little evidence to support the notion that the traditional student/professor mentor model of graduate student training (also a form of experiential learning) is ineffective at providing relevant job skills for our students. The general apathy of the industry sector towards engagement with the academic side of the skills development table would present the conclusion that they (industry) are reasonably happy with the skills that their future employees possess from their university training and that this push towards alternative, more applied research foci for universities is not based on sound evidence-based reasoning.

References

Canadian Federation of Students. (2013). *Public education for the public good.* Retrieved from: http://cfs-fcee.ca/wp-content/uploads/sites/2/2013/05/CFS-2011-Public_Education_for_the_Public_Good-EN.pdf

Dewey, J. (1938). *Experience and education*. New York: Simon and Schuster.

Doolittle, P. E., & Camp, W. G. (1999). Constructivism: The career and technical education perspective. *Journal of Vocational and Technical Education, 16*(1). Retrieved March 5, 2016, from https://ejournals.lib.vt.edu/index.php/JCTE/article/view/647/692.

Joos, B. (2012, January). NSERC's discovery Grant program: Disquieting changes & why they matter to Canadian science. *CAUT Bulletin, 59*(1). Retrieved from: http://www.cautbulletin.ca/en_article.asp?articleid=3385.

Kolb, D. A. (1984). *Experiential learning: Experience as the source of learning and development*. Upper Saddle River: Prentice Hall.

National Science and Engineering Research Council. (2013). *Government of Canada supports scientific discovery for thousands of researchers and students*. Retrieved from: http://www.nserc-crsng.gc.ca/NSERC-CRSNG/ProgramNewsDetails-NouvellesDesProgrammesDetails_eng.asp?ID=412

"The Map Is Not the Territory": Adding Value to Technical GIS Education

Mryka Hall-Beyer

Abstract GIS has developed by matching technological possibilities with problem-solving requirements. With each level of added complexity, the temptation is to specialize in the technology that is easy to explain to colleagues, to the detriment of breadth of knowledge with its ability to integrate GIS into larger systems. With only technical emphasis in GIS, one can mistake added data quantity and algorithm complexity ("map") for the application ("territory"). With added breadth of knowledge comes increased understanding of both systems and domains, plus increased ability to innovate, communicate and advance both the individual's economic usefulness and their public role. Undergraduate GIS training should establish outcomes that guide the student through increasing technical expertise, including the ability to speak effectively with non-GIS experts. This will provide degree holders with marketable skills for open-ended jobs and will be a valuable educational path for technical graduates looking to keep up with developments. Both graduates can continue to benefit from ongoing technical training offered in the workplace and through short courses. A broad GIS education adds value to any specialization that needs to examine spatial relationships. It allows making of increasingly sophisticated maps, without mistaking the technical output for the territory of the domain.

Keywords Education · Employment · GIS · Mapping · Undergraduate training

Introduction

The first GIS system was proposed in the early 1960s by the late Roger Tomlinson,[1] as a way to accumulate, file, retrieve and to some extent manipulate data that was linked to a specific geographical location. Its original application was to Canada's

[1] Like any complex non-patentable idea or invention, the origin of GIS is contested; in this case

M. Hall-Beyer (✉)
Department of Geography, University of Calgary, Calgary, AB, Canada
e-mail: mhallbey@ucalgary.ca

S. Balram, J. Boxall (eds.), *GIScience Teaching and Learning Perspectives*, Advances in Geographic Information Science,
https://doi.org/10.1007/978-3-030-06058-9_5

Land Inventory, for use by a government charged with managing multiple users over an extremely large and diverse land base. In its broadest sense, land managers (especially farmers) had been using GIS for centuries if not millennia, and certainly professional geographers used the concept before technology was invoked. Any time a phenomenon is noted on a map or even filed in a spatially conscious human brain, a geographical information system of sorts exists. Overlaying same-scale maps containing different data has repeatedly proved effective for developing new perspectives leading to scientific breakthroughs. A pre-GIS modern example is the reportedly serendipitous superposition of 1950s seafloor maps that underlay the development and validation of the plate tectonics paradigm in geology (for a popular account, see Felt 2012). Light tables and write-on transparencies could be said to form the first GIS technology.

Nevertheless, by the 1960s nascent computing systems begged for use in recording, storing and retrieving large amounts information. All it took was to systematize a way to link each data point to a location. Most of the thinking and information processing, once retrieved, would be performed by an expert in the field where the data was to be applied. As time went on, increasingly complex processing steps became feasible, with increasing requirements for expertise in the systems themselves: subject area expertise plus a glorified filing system no longer sufficed. This development posed a danger that the tail (the system) would wag the dog (the domains area). In other words, GIS became in danger of mistaking the maps produced for the territory as seen by the subject discipline. I will propose in this paper that by 2016 GIS had come round to a position in its spiral of development, where GIS professionals must emphasize their ability to tailor spatial tools to specific domains. Students need to be trained and educated to maintain the integrity of the process so that GIS is neither undersold nor oversold to their future employers and clients. At the classroom and program level, I propose that the hoary[2] concept that "The Map Is Not the Territory" is an overarching theme that can guide this discussion.

Most of the argument here derives from experience in teaching GIS per se and of teaching physical geography, cartography and especially remote sensing. Reflection on broad concepts has led to experimentation and winnowing of ideas about courses and programs serving students graduating at various levels into today's and the future's GIS job world. Program development experience has meant constant

---

by – among others – the Harvard University Laboratory for Computer Graphics (http://news.harvard.edu/-gazette/story/2011/10/the-invention-of-gis/). Stories grow up around inventions and discoveries as well. Tomlinson's particular claim is generally accepted, and the details are not important for our purposes in this paper. For those interested in backstories, see http://www.smithsonianmag.com/history/unlikely-history-origins-modern-maps-180951617/

[2]Wikipedia (http://en.wikipedia.org/wiki/Alfred_Korzybski) attributes this phrase to Alfred Korzybski (1879–1950), who was a pioneer in general semantics; however I have not been able to verify this fact. Its disciplinary or philosophical origin does not seem to hinder the phrase's usefulness in focusing geographical discussion.

revision to prepare students at various levels to meet the current job market needs as well as to advance GIS theory.

GIS: What Do We Mean Here?

The term GIS is used here in its original form of geographical information systems, a term linked inexorably to the computer science home of the concept. A "system" is a set of related parts that work together (Merriam-Webster 2016); it can be simple or complex. A GIS is a system that stores and manipulates pieces of information that are tagged by a location on Earth. The concept of spatial tagging has expanded beyond the strictly geographical to coordinate systems at other scales: one can apply concepts and procedures that involve spatial relationships to microscope slides or the insides of closets. The details of data storage and manipulation overlap mathematics (especially statistics and its offspring spatial statistics); the system would not be possible without computer science translating everything into and out of digital language. GIS thus depends on developments in these related fields and can contribute to those fields.

Like geography itself, GIS can be accused of "colonizing" other areas of expertise because it reserves the "right" to have something to say about anything that takes place "somewhere", i.e. in a defined location. Because of this, other terms have been used to expand the GIS specialty or to limit or subdivide it. Thus we have *Geomatics*,[3] which emphasizes the location rather than attributes attached to it and overlaps with the specific skills and legal certifications related to surveying. We have *Spatial Science*, which concentrates in analytical and statistical fields. It is perhaps too close to "Space Science" i.e. rockets and astronauts, for widespread adoption. A common expansion of the classical term GIS designates the S to mean "science". This emphasizes the particular specialized knowledge required to systematically investigate how spatial relationships among attributes help us understand the way the world works. *GIScience*, then, is focused on research and usually requires collaboration with a subject expert. Many jobs in GIS that go beyond simply finding a location will require such collaboration, whether in research or in projects.

---

[3]Geomatics was a term proposed in French in the early 1980s in *The Canadian Surveyor* (Paradis 1981). It has been used more in Canada than elsewhere, perhaps because of its bilingual origins and ease of use in two official languages. A Google search returned only one non-Canadian link within the top 50 hits. The origins of the term within the surveying community, however, have (perhaps unfairly) led to a connotation that Geomatics is most properly concerned with location itself, especially with high location precision and accuracy, rather than with attribute analysis and modelling. The two are closely related and in the author's opinion should not be artificially separated by terminology. Within a GIS, location is empty without attributes, and attributes are orphaned without location.

In the public arena and "party conversation", I have found that GIS is often mistaken for GPS (Global Positioning System). This is ironic, because what the public takes to be GPS is in fact a GIS system: few car drivers would be content with a system that tells them their location without GIS data and functions telling them how to get where they are going. But the everyday usage of "GPS" now seems to be what specialists called a GIS a few decades ago.

Nevertheless, despite some terminological fragmentation, GIS remains the common term for the ensemble of systems, functions and output related to location-based services. Online searches for software, services, data, conferences or education using GIS as a keyword return hits from every part of the world in multiple languages.

A Brief History of GIS Developments

GIS domains of applications and the skills required to use the system have changed rapidly. At the same time that function complexity has become possible, tools that package the more straightforward processes (e.g. distance measurement and route tracing) and general map interactivity have also become widely available to the public. Thus the earlier-developing and simpler GIS tasks no longer require a specialist but can be operated in an elementary classroom or largely self-taught by adults. As computing and networking capabilities have become more complex and ubiquitous, large numbers of people find themselves using GIS tools and concepts without any training in them whatsoever. Few people think they are using spatial analysis when they activate their dashboard GPS: either they rely on their own "intuition" about space or simply accept without demur the GIS-delivered answers or maps. The stories of people blindly following their GPS/GIS instructions into dangerous situations are too real, and some have by now wandered into the realm of urban myth.[4] This broad-stroke picture sketches a mounting spiral of technical development, use by domain experts, popularization, further development and so on. Paralleling it is an evolution of GIS teaching at various educational levels. This section details some of these co-developments.

GIS must depend on computer hardware and operating systems and on the advances in "big data". GIS's increasing needs also drive developments in these areas, particularly software. The first GIS was developed for government, largely for natural resource information management. This joined the needs of census reporting to develop a topological data structure for ease of handling and automated mapping (Deakin 2002). Both governments and research institutions were large enough to have access to expensive mainframe computers of the 1960s and 1970s. They also covered enough territory to benefit from, and soon require, a locational indexing system. With a large-institution proof of concept, other organi-

[4] See http://www.npr.org/2011/07/26/137646147/the-gps-a-fatally-misleading-travel-companion for a media example and commentary from 2011.

zations with similar but more restrained needs and resources became interested. These would be research laboratories in universities and large companies, again especially those dealing with land resources. These groups collaborated with commercial enterprises to develop GIS through the heavy lifting phase of figuring out how to format and store data. They shepherded it through the early frustrations with interoperability and standardized data formats. Teaching was undertaken as research-level apprenticeships attracting students from both technical and applications domains. Little ordinary undergraduate GIS teaching was on the horizon.

The rapid improvement of computing speed and storage space through the 1970s and 1980s laid the groundwork for the personal computer to become ubiquitous. The simple introduction of low-cost colour monitors opened up many previously unthinkable applications. Yet even in 1997 the average user was expected to be content with mere 16 colours (Star et al. 1997 and 2011; Hall-Beyer 2012). Despite lingering limitations, the spread of the personal computer permitted smaller institutions and the tech-savvy public to adopt basic GIS.

As GIS came within the reach of the undergraduate student, instruction in its use was also broadly conceptual, relying on the common sense approach. In 1987, I taught "GIS" concepts at an intermediate undergraduate level using a series of plastic transparencies and a pin. At this level, it was not much more sophisticated than asking the same students to look at atlases, make notes and then think their way through to a map generalized for the question they were asking. The immediacy of seeing the overlaid data inspired a new level of awareness of things the students might not have thought about without a visual aid. After having participated in such an exercise, students were very amenable to the idea that "The Map Is Not the Territory" since they were keenly aware of having exercised selection and generalization in making the overlay layers and then in extracting hypotheses from them.

By the early 1990s, the same university had acquired a few desktop computers with colour monitors and added some specifically GIS software. There was no local data available in a usable format. A flood occurred on the campus, and immediately the local public became interested in how our new software could shed any light on future risk. To answer that, students surveyed elevation points (without GPS) and interpolated a surface, hand-digitized building and facility outlines and learned the GIS functions allowing them to fill with water to certain elevation levels. Here again, their own experience of having neither tools nor training enough to create precise and accurate data made them acutely aware of the uncertainties in their analyses. What to say when a neighbour asked if their house would be at risk in the next flood: yes or no? Through experiential learning using real but flawed data, students became aware of the nature and requirements of hydrological models and also of the nature of the maps they created using the GIS technology. No, The Map Was Not the Territory: but there were glimmers of hope that it might become so. The watchword was "if only we had more and better data and faster computers and better map plotters".

Undergraduate students were not the only people using that watchword in the 1980s. Data acquisition, validation and cost became a major issue not only for the science user but for the everyday reader of output maps. A GIS that had been born from a need to file and retrieve large amounts of existing data increasingly drove the acquisition of data. Should government data be public? In what form? If not public, what people should have it and how should they be trained? Would this create many new high-value jobs (the hope of the students), or would it abolish existing jobs (the fear of those already employed in traditional mapping fields)? At this time, the "geographical" in GIS determined that a great deal of basic GIS training would be housed in geography courses, with the computer technology side in the companies and of course in specialized university research departments. Surveying applications were quite separate: this is the era when "geomatics" was born as a term. The typical course taught GIS as a way to overlay data to answer a question formulated by geographers. In practice, this meant mostly physical geographers as they were more comfortable with quantification.[5]

As can be seen from these anecdotes, the concepts required to acquire data and manipulate it to make simple maps were well within the realm of what could be self-taught with access to minimal software and consumer hardware. This being the case, almost any geography program could include a section on GIS within a cartography course or if resources allowed have a stand-alone GIS course and laboratory. Google acquired Keyhole Technologies in 2004 (Cowley 2004), leading to the release of Google Earth the following year. At this point enough data were freely available that anyone interested could become as much a GIS user as the student in a university geography classroom in the late 1980s. Instruction in the use of Google Earth is now in the elementary schools and public libraries, not in the geography classroom. Few of these new users had direct experience of the uncertainties of acquiring their own data and building their own systems. It did not occur to many people to address these potential problems, unless they observed some image anomaly like a one-winged airliner.[6] Most of these anomalies have long since disappeared from the web, leading to even fewer user questions about outcome quality.

As computers increased speed and storage capacity at a dizzying rate in the 1990s, and as finer-resolution colour screens and graphical interfaces appeared on the typical desk, many more post-secondary institutions acquired computer labs and introduced explicit courses dealing with GIS, increasingly interdependent with digital cartography. Much more low or no-cost data began to be available through public institutions. This was often of very high quality and contained a good deal of metadata. Students could select from existing data and even begin to evaluate it for suitability, rather than tailoring projects to what was available. Key data to allow labs were digital elevation models (DEMs), road networks, natural features such as hydrology and extensive satellite and airborne imagery, and census or other

[5] This is not to make a value judgement, as of course geographers concerned mainly with human phenomena have adapted GIS to their uses from the start, as explained by the involvement of census needs in early GIS topology definitions. However it remains true that the earlier adopters were, not surprisingly, concerned with land resources, since GIS had started for them and co-evolved with their needs.

[6] http://www.geekabout.com/2008-01-03-440/top-17-most-bizarre-sights-on-google-earth.html includes both data errors and common misinterpretations due to such things as image viewing angles.

administrative boundaries. Using GIS to its capacity began to require knowledge of statistical manipulations and modelling. A simple introduction to GIS as part of another course was no longer adequate, and a second course could be justified.

Also in the 1990s, students outside geography began to take an interest in what GIS might have to offer their field of endeavour. Engineering, computer science and mathematics, beyond surveying, assumed an important role in the emerging GIS expertise complex. Independent stand-alone courses appeared at smaller institutions, and technical institutions introduced GIS programs.[7] About the same time it began to be expected that all geography students, of whatever specialty or of no specialty, should have at least an introduction to geographical methods, now taken to include not only GIS but statistics and remote sensing as a science distinct from raster GIS.

There have been numerous attempts, at first led by existing professional bodies such as the ASPRS,[8] to systematize and certify knowledge surrounding GIS and skills in its use. These seem to have been particularly useful when GIS was being added to the repertoire of skills demanded of professionals already having certification (such as surveying engineering) in an established field with legally mandated licensing in many jurisdictions. A bit later, the wide use of GIS within "homeland security" and other military disciplines (including disaster relief) led to similar demands for a uniform certification and skill set definition (USGIF 2016). Although remote sensing and GIS have become important fields in Canada for professionals such as geologists (e.g. APEGA[9]) and surveyors (e.g. CBEPS[10]), to the author's knowledge, there is no particular push for GIS to have its own certification in Canada, though individual organizations (e.g. URISA[11]) may offer it for those who would find it advantageous within their own careers (Murphy 2013).

GIS was still a novelty, but becoming less so. Into the early 2000s, there remained both a "scare factor" for the non-quantitatively oriented student and a "cool factor" for the technically oriented one. A good deal of jockeying occurred about what level of training should be given, the degree of specialization required or possible, the level (year) of specialized courses and the creation of specific credentials to follow the student into the job world. Interest was expressed in completely stand-alone training programs outside of universities, such as at technical institutes, tech-

[7] http://gisgeography.com/college-gis-certificate-programs-list-canada/ maintains what it calls an "exhaustive list" of college (preuniversity and technical bachelor's) level GIS programs in Canada. This shows the extent of current offerings but is presented in a primarily advising/marketing format and does not mention the year of inception of each program. Those the author is directly familiar with became available in the early 2000s.

[8] American Society for Photogrammetry and Remote Sensing: http://www.asprs.org/Certification-Program.html

[9] Association of Professional Geologists and Engineers of Alberta: https://www.apega.ca/apply/exams/technical/courses/geomatics/

[10] Canadian Board of Examiners for Professional Surveyors: https://www.cbeps-cceag.ca/sites/default/files/C%205%20Geospatial%20Information%20Systems%20Study%20Guide_0.pdf

[11] Urban and Regional Information Systems Association: http://www.urisa.org/careers/gis-professional-certification/

nical programs in high schools and for-profit training and certification. Because of the highly technical nature of the file formats and algorithms, it became possible to defend a GIS program that in fact had almost nothing to do with the traditional concerns of geography, which I will simplify as "where is it, and why does location matter". Simply "where is it" and "what buttons do I push" could occupy several courses, which might leave "why does it matter" quite out of the picture, or at least in the realm of another specialty not concerned with producing the map or other output. The Map threatened to overtake The Territory being examined.

One of the temporary casualties of this emphasis on GIS was a split between classical cartography and the new technologies. When maps were hard to make, when excellent maps were works of high art and craft, the selection and arrangement of map elements had to be guided by a deep knowledge of both the domain to be shown and the generalization level required. With maps being easily produced by GIS, it is easy to devote the classroom and lab time to the production, rather than the planning and critique, of cartographic output. The cost to a student of not thinking too clearly about things like abstraction, scale and generalization became lower, as corrections could easily be made. While the concept that "The Map Is Not the Territory" remained honoured and even emphasized in the curriculum, there seemed to be a lot more territory going into the production of the map, and it was ever more possible to sublet the selection to the technology and to whomever created its default values. Questions like "why did you choose to divide your classes at 234.567 and 654.321 instead of, say, 250 and 650?" were less deeply considered. In some cases they were not considered at all, so long as the map "looked good".

In everyday life, technology exited the realm of geekdom. The earlier need and opportunity to personalize computers to one's own needs and desires were replaced by convenience. The technology consumer market focused on people not desiring personalized interfaces and functions but rather desiring to be connected and mobile. The social network was born at about the same time as Google Earth. While this public shift was going on, of course the GIS world was also developing rapidly. Software was made simpler and more intuitive; algorithms were sometimes packaged in "wizards" or later "apps" that only required selection of data inputs and parameters and contained a convenient default set. Software became less expensive, especially to educational institutions including K-12 schools. Network connections spread and accelerated. Above all, data became easier to use. This spilled over into some students' attitudes that whatever the computer said must be right and if the output disagreed with "common sense", then either the computer was mildly diabolical or more likely correct.

At the same time, people with no expressed interest in "technology" were using, and in fact relying on, computers for everyday efficiency. One could choose between passively blaming an algorithm that didn't work and buying an "updated more user-friendly version". Learning *about* computers, and actually being able to make one work, became thought of as a specialized job skill, not a tool to help whatever one might want to do. This can slide into a complaint when "The Territory Is Not the Map". Rather than finding a map faulty if it lacks something known to be there, one might think that the territory – reality – is at fault if it fails to conform to the map. A new area needed to be taught at a less advanced level: the nature of uncertainty in maps, with increasing emphasis on theoretical as well as practical problems of generalization.

For the beginner, confusing Map and Territory was partly justifiable. Increasingly, software capability, computer speed and the readily available highly detailed data for wide swaths of Earth's surface produced maps that looked like the fount of knowledge.[12] Anyone could zoom out to see the whole Earth at once or zoom in to see their house. Places without houses were of less interest, so few people noticed the lack of detail in places where high-resolution imagery was not available. With time, these vaguer places became fewer. With this overwhelming amount of available data and intuitive rendering, the need for technical knowledge appeared less obvious. And, without technical knowledge, the concept of metadata, deliberately acquiring or selecting data for a particular purpose or meaning and validation, became foreign.

For those entering school with this background, there is a strong temptation to want either "what buttons do I push" or "just show me the picture/map" courses and labs. At the same time, there is increasing pressure on the educational system at large, including of course the university system, to produce graduates able to fill job vacancies, or to become tech entrepreneurs, immediately upon graduation. Advanced education is also becoming more expensive to the individual student, as well as to the public at large through taxes and endowments. Both of these tendencies join to favour a bifurcation of GIS education into the technical and the general, and an assumption that the jobs go to the former.

If at the same time education is seen as more cost-efficient when mass-produced and strictly technical subjects can, in theory, be more easily conveyed through online training than in a classroom, then webinars might be believed to adequately train students for the GIS job market. This training is an excellent way to learn the ins and outs of algorithms but is less effective at inculcating critical analysis. Webinars are often linked to specific software, and so cannot avoid a quasi-marketing aspect. Rapid results may be favoured over validation. Part of this attitude comes from a public-relations failure, where GIS is seen as one more e-appliance that anyone can buy and use. Part of it comes from a student bias for purely technical training so they might seize entry-level jobs requiring up-to-the-minute software skills but little in the way of theory or analysis. Thus maintaining a critical and experimental education at the university level becomes difficult: a facet of the ongoing debate about the nature of university education (Blouw 2013).

I would maintain that an intermediate position is most appropriate. Recognizing that newly hired personnel in many GIS-related jobs will need analytical and

[12] While writing this chapter, I was contacted by a reporter wondering why the new bus arrival times for the city's transit system were so often wrong (by a whole minute sometimes!) and thinking it was some fault in the "GPS" system. It took a while to explain that the fault, or rather uncertainty, was likely in the city's traffic model that joined the location of the bus from GPS, the location of the call, and forecast the travel time using route speed limits. The reporter echoed a common impression that GPS could "see" everything in the future rather than model and predict it with attached uncertainties.

self-teaching skills to advance in jobs, technical schools themselves promote their graduates as having higher skills than simply pushing buttons as directed. In some cases, excellent average technical graduate results may be influenced by their graduates also possessing prior university degrees and workplace experience. For these, their tech skills are value added and put them easily in line for job advancement. What they have added to the technology is general knowledge from other courses to support their technical ability in a GIS lab. GIS has now advanced far enough in theory and capacity that a GIS analyst, able to plan and direct projects, needs dedicated courses that include critical thinking, evaluation and self-learning skills. Otherwise any entry-level jobs will be automated leaving the tech-alone student stranded. For example, just as there is a difference between being able to install a plumbing system and being able to design one, there is a difference between being able to accurately update a survey plot and being able to survey it and a difference between being able to make a system that will allow information queries and one that will produce a map of wildlife habitat under various proposed planning scenarios. As technical schools recognize this, they become closer to universities (e.g. applied bachelor's degrees) but less accessible to students explicitly choosing not to study at a university level. This is not to denigrate the various levels of GIS expertise taught in different places but rather to attempt to discern the specific contribution to the field required of universities, in particular university geography-taught GIS.

This GIS development cycle seems destined to continue: automation of more established methods (fewer entry-level jobs) and increasing need to develop new, more complex procedures to support research and innovation. Basic GIS capability – making a simple interactive map, for example – is expected of any geography graduate but also of graduates in other fields that look at things spatially. Maps are being more and more needed for public presentations, book illustrations, stakeholder sessions and legal briefs, for example. Similarly there is a sweet spot in the K-12 level for teaching GIS skills that everyone can benefit from in daily life and work.

It sounds like geography as a discipline may be in danger of losing the G in GIS. If geography programs turn too technical, they may be ceding the territory to disciplines in danger of thinking that "The Map Is the Territory". I suggest that geography can add distinct value to technological competence by insisting on a broader scope of GIS. To do this, GIS has to specifically take domains of application and strong communication under its wing.

Whither GIS Education? The Range of Job Prospects

Technical GIS education, with the emphasis on the "S" in GIS, is attractive to students where it promises immediate employment in large GIS departments of industry and operational units of government. In these places, there is likely to be a GIS department with supervisory personnel both experienced and knowing some breadth in GIS. The entry-level, technically trained graduate then has a limited opportunity for career advancement. They may choose further job-level training, and essentially teach themselves, keeping abreast of the evolving algorithm and apps scene. This is certainly possible, though it requires independent initiative and may be acquired just-in-time. The breadth of perspective to innovate may be lacking, and credentialing for job mobility is less straightforward than through educational institutions.

Alternatively, and probably more efficiently, the person who wants to advance in the technical side of GIS might choose to obtain further education at the university level concentrating in the engineering or computing sciences. Because of high cost and competitive entry into these programs, the technology graduate best positioned for enhancing job market skills in this way is one who has previous university experience and a good technical school record. The possibility for part-time and online education is very attractive for this group of people.

In the previous scenario, the student who has chosen to forego university (or an applied bachelor's degree) for time, financial or aptitude reasons may be disadvantaged in future promotion within GIS. A major advantage of technical degrees is their up-to-date software teaching, producing a graduate who can slip into a job requiring use of current procedures. This technical diploma graduate may find jobs harder to qualify for in smaller companies and institutions where they are the only GIS person around. Even explaining what GIS can and cannot do, or what pitfalls to avoid in a given situation, may not be within their capacity. For this kind of solitary or more responsible job, additional training is necessary that will give a breadth of outlook and an enhanced ability to make informed judgements and evaluate the results. The ability to function in this situation, as well as to advance in the GIS department of a larger workplace, needs to be taught at the university level if it is to mesh with technical GIS training. Students taking introductory courses need to start right out in this pathway, and the objective needs to continue to guide course content at more advanced levels.

Whither GIS Education 2: Outcomes and Assessment to Achieve These Goals

A university-level undergraduate GIS education needs to prepare students to become GIS analysts and in turn to benefit from the full university education. This means graduates should, with minimal job experience, be able to assume project oversight and responsibility for new GIS initiatives. They may assume complete responsibility for GIS in small organizations. Of course university students need to acquire technical skills, but they need to be constantly referred back the ability to plan, execute, work in teams and evaluate and communicate their results. Critically, they need to be aware of many domains and their particular needs. An ability to communicate in GIS language and translate it into other specialist languages will be a great asset to the GIS graduate, so this should have its place in their curriculum.

An increasingly adopted way to ascertain that entire programs of study achieve the desired results is to use "educational outcomes" as a framework. Although widely applied the field of health education at the university level (Davis 2003), and originally popular during the 1980s and 1990s, their current applicability to technical fields such as GIS is immediately apparent. Outcomes specify what a student can do, and GIS involves doing. Outcomes ideally express what students can *do* upon graduation, not only what they *know* or *understand*. Each course within a program may have its own more specific set of outcomes, aligned with the higher-level ones. Naturally, effective "doing" will involve a great deal of knowledge and understanding, but it must be in service of the "ability to do". An advantage of expressing course and program outcomes in the active voice is ease of structuring assessment to test if they in fact do what they say they can do. Using program level outcomes is a reminder that for a complex skill like GIS, practice of the same skill many need to be repeated at increasingly higher levels to cement the original exposure to the skill or concept. Finally, cleaving close to outcomes provides students with ready-made talking points at the job interview and products to show.

GIS Outcomes at the University Level: A Proposal

"The Map Is Not the Territory" works as an overarching theme. A GIS education that wants to make sure the higher skills are achieved might start with a set of outcomes such as the following. These points are broadly based on the University of Calgary Geospatial Sciences undergraduate stream within the Geography Department, with additions specific to the purposes of this chapter:

- Acquire, evaluate, format and integrate data appropriate to a GIS project.
- Choose and execute algorithms using commonly used software to answer simple questions.[13]
- Link algorithms in a logical sequence to answer more complex questions. This includes being able to plan different conceptual routes to the same outcome.
- Chart interrelationships among many sub-sections of a project.
- Carry out a complete evaluation of the project outcome, from the point of view of uncertainty tracing as well as congruence with information taken as correct by the application area (domain).
- Create comprehensive and comprehensible output in text, graphic and cartographic format for a chosen audience.
- Carry out independent projects from start to finish, including evaluation of strengths and weaknesses, compromises made with the realities of data and timelines and meeting objectives of an organization or domain.

[13] I do not use the GIS term "query" here since it tends to cloud the issue for many students in my class. I limit "query" to the technical sense of point selection according to user criteria http://support.esri.com/en/knowledgebase/GISDictionary/term/definition%20query). I use the terms in this way purely to minimize misunderstanding in a GIS classroom.

- Communicate effectively with nonspecialists in a few chosen domains, for example, urban planning, engineering, environmental assessment, health services provision and municipal planning.

For each outcome, there is an introductory way to get there without reducing activities to recipes. For a student being introduced to geospatial concepts at the level needed by any geographer or geoscientist, simple examples need to be presented. The main purpose at this introductory level is to produce interest and above all to reassure students who may be intimidated by the technical end, especially if they have not previously used software requiring choices of algorithm or input parameters. This works well if exercises are set up so that easily correctable mistakes will be made. Students can themselves collect location-based data through learning to use a GPS and importing a spreadsheet into a geodatabase. At the same time, more complex tasks (such as reprojection) can be introduced in theoretical sessions, whether it be through lecture, directed discussion, assigned reading, production of a module to teach children or other method. Maps can be critiqued, often with humour.

With this basis, each successively higher level can reinforce existing competencies and introduce a greater level of independence. At the same time, assignments can move from providing all data for an exercise (the student will assume it has all been corrected and approved by "an expert") to having the student become responsible for acquiring and evaluating it, and performing data cleaning as required. This can start with direction to a validated data site, so that questions are minimal, and the emphasis is on searches and downloads and imports into software as required. At a higher level, it is a great advantage for students to gather their own data in the "field" (whatever that may mean for the domain) for some labs or projects. Students themselves would plan, organize and record the information, as well as select what data to record and how it is to be quantified or noted. Data collection provides a strong touchstone for theoretical discussion of concepts of scale, precision and generalization.

The Importance of Targeted Communication Within Domains of Application

The last and most integrative outcome in the list above emphasizes communication to various audiences and at chosen levels of technical detail. This is the area where a broad background in the student's program becomes most advantageous. It is also where students gain important skills that will be necessary to advance to higher levels of responsibility. These might be considered "soft skills" but if so they are concretely targeted at areas that use GIS. A very important benefit is to learn the technical vocabulary of the domain and how GIS vocabulary might best translate. A specific methodology that constrains GIS use might also be learned, to better work through appropriate GIS procedures. An example would be becoming comfortable with the precision and scale required for archaeological work. Since different students will gravitate to different domains, all students in the class can become aware of the diversity of potential domains areas. Since early GIS co-developed with physical geography and census-based data, exposure to the developing possibilities of GIS in "digital humanities" (e.g. "HGIS" (Historical GIS): Bonnell and Fortin 2014). This will broaden the student's knowledge of possibilities as well as provide grounding in a particular domain of their choice.

In the early years of GIS education, when it was not yet a regular part of the curriculum, students actually choosing GIS usually already had a domain where they wanted to apply GIS. Archaeologists came to GIS courses wanting to apply GIS to archaeology. So did planners, wildlife ecologists, surveyors and computer scientists. Today, university GIS education is increasingly situated in the ordinary progression through an undergraduate degree program, so this pre-existing interest and expertise is less universal. GIS therefore becomes only one part of a wider disciplinary competence. Students self-select for a stream emphasizing engineering/surveying/sensors or one emphasizing applications/domains focus, when they choose their overall degree program. The two do not overlap extensively and usually mean opting for engineering or for geography programs. These "wings" of GIS education interpenetrate at more advanced levels, but the undergraduate so far seems required to choose one or the other, because of prerequisite structures and preuniversity choices particularly in maths – and student overall interest, of course.

Within the applications/domains streams, however, a wide breadth of interests can profit from a diversity of skills and experience. Students can be given assignments depending on multidisciplinary teamwork, requiring presentations with a given audience in mind, and encouraged to build portfolios of their projects aimed at future workplaces. Domain experts, not just GIS professionals, can be invited to the classroom and students brainstorming potential application of GIS to this area. These are all possible approaches to ensuring the desired student outcomes: in practice it will not be possible to integrate all of them into every level of a GIS stream.

But Will the Domain-Oriented Graduate Have the Necessary Technical Skills?

Tight economic situations lead to doubts about the advantage of a university degree for a coveted entry-level job, as opposed to a technical certificate or diploma (or an applied degree). That advantage should be in subject breadth, depth and self-directed learning and creativity in solving problems. Ideally, in a rapidly changing field like GIS, it might include an appreciation of the short-term directions of the subject. At the most basic level, a single introductory undergraduate GIS course should prepare a student to make judicious use of spatial data within a chosen domain. If this is what an employer wants, the university graduate should be able to compete with a graduate with similar domain background but who lacks basic GIS skills, provided the applicant can articulate the potential GIS contribution. For specifically GIS jobs, to get in the door, students must be able to demonstrate basic data-handling competence and use of common software. They will have extensive experience in only a selection of the most recent highly specialized software. They should demonstrate (through portfolios or more traditional CV lines) that they master the whole sweep of projects from conception through communication. At a higher job level, more advanced education will be necessary, probably integrating the domains and technical aspects and including some programming or app development. This level is not the object of this chapter.

K-12 GIS Education: Preparing Future Students for Any Level and Inciting Interest

Before arriving in a technical or university program initiating and developing GIS skills, students need to know that such a thing exists. Domains are very important here: GIS in the K-12 schools can readily support inquiry and experiential learning across the curriculum. Widespread online data and tools can be incorporated into elementary and secondary classrooms if teachers are aware of it. I have mentored a course for advanced students in Geography (including Urban Studies) and Geology, where students act as subject experts alongside elementary teachers for a semester. Students in this program report believing that one of their most significant contributions is making teachers aware that GIS is available and how easy it is to use. Along with enriching the classroom, the legacy they leave is often increased awareness of how spatial information fits into so many niches in government-mandated programs of study, even if they bear the name "science" or "social studies" rather than anything containing the phoneme "geo".

This is not the place to examine the extensive literature about children and spatial concepts (see Moorman, this volume, and Mohun and Mohun 2014), which would be important in any K-12 curriculum review. Post-secondary GIS educators will likely be taking in undergraduates possessing a wide variety of spatial awareness and background, from zero to "geek". University-level courses will have to introduce concepts from the beginning, assuming little. However it is to our advantage to encourage formal and informal inclusion of spatial concepts at earlier education levels, to make the possibilities of GIS more widely known to incoming students and also to their parents and to education policy makers and potential employers. A bad scenario would be to allow everything spatial to default to unexamined use of consumer electronics and websites.

Back to "The Map Is Not the Territory"

The approach to GIS education suggested here insists that students will be most effective in the workplace if they combine a substantial amount of technical know-how with a good deal of critical reflection on what they are doing, practical experience in self-learning of different possible ways (both technical and conceptual) to solve spatial questions and a growing ability to communicate across disciplines or domains. This may limit the variety and immediacy of software knowledge and programming skills, but it maximizes the potential for future self-learning, creativity in unfamiliar domains and ability to adapt to a changing technological environment. This structure makes it attractive for technical graduates to upgrade their skills to the university level and can help employers differentiate the various levels of GIS expertise needed for different jobs.

"The Map Is Not the Territory" means that reality always has more about it than we can capture in our data systems and more than is dreamed of in our analysis and output. GIS is able to handle a great deal of data and complex tools, to the extent that it is tempting to mistake our output for the sum of reality or to think we have captured all that "really matters". GIS above all fields is able to provide unique insights into whatever is concrete "out there" by adding unique spatial dimensions to every domain. It is able to solve many practical problems, or ameliorate others, and to contribute a great deal to necessary wisdom in making choices. It is not in the business of changing "The Territory" to match "The Map". It is important to our future GIS specialists that their ongoing education emphasize this fact, to maximize their potential contribution both to the economy and to the increasingly global dimension of citizenship.

Acknowledgements I would like to thank several people who have supported my teaching experimentation and critical thinking about practice. Dr. Greg McDermid of the University of Calgary strongly encouraged me to begin teaching outside of my more comfortable areas of remote sensing and associated raster GIS and to dive into the much broader and less wieldy areas of GIS in general. Dr. Lynn Moorman of Mount Royal University engaged in many years of discussions and co-teaching, starting with a K-12 teacher training course "way back" in the 1990s before it even seemed possible. Don McLaughlin of the Calgary Separate School District has been both a colleague in bringing GIS to elementary students and a GIS student in my courses and student project mentor. Finally, thanks go to a succession of Department Heads in the University of Calgary Department of Geography who fully supported teaching experimentation and participation in teaching and learning-enhancement programs, even when it limited my time for other important things.

References

Blouw, M. (2013). Universities should educate: Employers should train. *The Globe and Mail*. Editorial Sept. 3, 2013.

Bonnell, J., & Fortin, M. (2014). *Historical GIS research in Canada*. Calgary: University of Calgary Press. 344 pp.

Cowley, S. (2004). Google buys digital mapping company. *PC World*. http://www.pcworld.com/article/118352/article.html

Davis, M. L. (2003). Outcome-based education. *Journal of Veterinary Medical Education, 30*(3), 227–232.

Deakin, A. K. (2002). Evolution of geographical information and visualization systems. *Encyclopedia of life support systems vol. II: Environmental monitoring*. UNESCO. Not paginated.

Felt, H. (2012). *Soundings: The story of the woman who mapped the ocean floor*. New York: Henry Holt.

Hall-Beyer, M. (2012). Book review of Star, Estes and McGwire 1997/2011. *Geomatica, 66*(1), 62–63.

Merriam-Webster (2016) https://www.merriam-webster.com/dictionary/system

Mohun, A., & Mohun, L. (2014). Spatial thinking about maps: Development of concepts and skills across the early years. *National Geographic Education Foundation.* http://education.nationalgeographic.com/education/media/spatial-thinking-about-maps/?ar_a=1

Murphy, J. (2013, February 5). GIS Certificaiton: Is it worth it? *Geomatics Canada Magazine*. http://www.gogeomatics.ca/magazine/gis-certification-is-it-worth-it.htm

Paradis, M. (1981). De l'Arpentage à la Géomatique. *The Canadian Surveyor/Le Géomètre Canadien, 35*(3), 262–268.

Star, J. L., Estes, J. E., & McGwire, K. C. (2011, unaltered reprint of 1997). *Integration of geographic information systems and remote sensing*. Cambridge: Cambridge University Press. 248 pp.

USGIF (United States Geospatial Intelligence Foundation). (2016). *GEOINT essential body of knowledge*. http://usgif.org/system/uploads/4022/original/EBK.pdf

Using an Online Format to Teach Graduate-Level Remote Sensing Basics

Mryka Hall-Beyer

Abstract A small-enrolment graduate-level remote sensing course moved online in 2010, to solve timetabling problems without sacrificing preparation of students for an advanced seminar course. We also wanted qualitative indication of any change in resource and time commitment. The course was presented 8 times, serving 100 students. Two courseware systems and two laboratory software packages were used; upgrading for organization and content was continuous. Student outcomes are equivalent to the traditional course. Advantages include asynchronicity and personal agenda control. The course framework also permits students to complete the course as an independent study without synchronizing with the primary course. A modified flipped laboratory has been tried. Costs are in technical and administrative support. Most online course support assumes large lecture sections. This course's success tests possible future directions at the institutional level. Learning Objects for advanced topics serve a small group of instructors linked by common institution or professional contact. Their relevance and wider availability await effective governance and accessibility of learning object repositories (LORs). It is not yet clear if there is a need for LORs for advanced technical material or if specialized webinars will better fulfill this function in a discipline heavily interdependent with major software development.

Keywords Graduate education · Learning object repositories · Online learning · Remote sensing

Introduction

Since the 1990s, software and hardware have been developed and implemented to streamline instructor-student communication and provide a constantly available repository for course materials. At first this simply replaced paper handouts and

M. Hall-Beyer (✉)
Department of Geography, University of Calgary, Calgary, AB, Canada
e-mail: mhallbey@ucalgary.ca

S. Balram, J. Boxall (eds.), *GIScience Teaching and Learning Perspectives*, Advances in Geographic Information Science,
https://doi.org/10.1007/978-3-030-06058-9_6

centralized individual at will private access to grades. Online was viewed as a particularly positive development for courses where a need was seen for additional forms of student-instructor and student-student interaction. With the almost universal availability of high-bandwidth connections and the explosive popularity of mobile devices (Chen and Denoyelles 2013), large volumes of material (including assignment submissions and lab data transfer) could be handled. Social media or "web 2.0" became integrated into learning management systems to allow extensive interaction with the instructor(s) and among students (Tess 2013). Entirely online programmes of all subjects became possible, and experiments were started with very large courses (MOOCs and others: Liyanagunawardena et al. 2013). Uptake was rapid for technical courses with large enrolments, where individual training aspects could be favoured. In the GIS field, ESRI developed focused, technical online courses (2016 version: http://www.esri.com/training/main/foundational) promising – or threatening – to replace the individually developed laboratory exercises for stand-alone courses. In early days less focus was put on advanced, small class-oriented courses or on those teaching and evaluating critical thinking. Online was more envisioned as a way to get some of the advantages of these courses into a larger format course. Given the high planning, design and startup costs in time and technology, small courses were not emphasized in converting to online formats.

How the course reported on here, "603", fits into these trends will be discussed below after the course and its purposes are described.

What Is 603 and Why It Went Online

The programme leading to a Masters of geographical information systems/science (MGIS) has been offered at the University of Calgary, Canada, since 1999, and has graduated well over 100 students. Before going online, 603 was a single-instructor course within the MGIS programme with minor technical and lab assistant support. The programme prepares graduates to seek work at both the analyst and management level of GIS, so critical thinking, problem solving and adaptation to meet the needs of a wide variety of application areas (such as forestry, geological mapping, risk analysis and planning, health management and urban planning) are the main foci of all courses. All students must demonstrate advanced competence in three core areas: GIS itself, remote sensing, and spatial statistics. There is also a major independent project providing credit for two courses. Five additional courses are required to complete the programme, which is capped by a project and oral examination. At the programme's inception in 1999, enrollees came from the job world, usually having some GIS experience but generally little broad knowledge or systematic training of the other core areas. To bring everyone to the same level, three additional "Basics and Beyond" courses were made available, one in each core area. Each is roughly equivalent to two undergraduate lab-based courses. Since 1999, increasing numbers of undergraduate programmes offer training in these areas, so more students enter the MGIS programme with some systematic study. However many new students are not confident of their remote sensing background or believe that their skills are out of date and so join the remote sensing neophytes in the course. This situation requires some flexibility in the curriculum content and favours a structure allowing students to choose among more advanced topics.

The author had delivered this course several times before deciding to take it to an online format. Within the programme, the online 603 was originally an experiment to see how students reacted. Could programme objectives be fulfilled? The idea was not "to have an online course" but rather to explore the ability of an online delivery format to accomplish and enhance existing objectives.

Within the course itself, online could offer several potential improvements. Most attractive was asynchronicity: no student would be required to be at a particular location at a particular time and would only rarely be required to be online at the same time as other students. Many MGIS students continue to work full- or part-time during their programme. Given the Geography Department home of the programme, many of these jobs involve field work away from the city for varying periods of time. We also accept international students, for whom enrolling is not only a major investment of time and money but a leap into a new culture and an "adventure" in immigration paperwork. Completion of one programme course online, potentially from home, might save their resources or allow an earlier programme start. Finally, if the experiment worked, then even local students or those not working inconvenient hours would be able to better juggle other academic, family and personal commitments. The course instructor could devote less time to lecture delivery and more time to student interaction, monitoring and incrementally improving the course.

Potential Risks

The risks in this move online were forecast to include the following: some students might have a hard time organizing their coursework and meeting deadlines; whole groups of students might have severe reservations about online learning; students with low computer bandwidth, particularly in international or remote field locations, might be left out; and there was a potential for unproductive discouragement as students attempted to acquire, install and learn to use the highly specialized image processing software required for the course. A complex structure would be needed to weave together all components of the course, and we wondered if students would have more difficulty in finding components than in traditional classroom and laboratory delivery. Pedagogically, we were encouraged by the developing online experimentation in fields where explicitly fostering student critical thinking and interaction are emphasized over facts and procedures (e.g. Moore and Simon 2015). The need for self-instructing labs was a particularly thorny potential problem.

Course Format and Structure

The course has been given every fall semester from 2010 through 2017 (eight offerings). One hundred students have enrolled. Of these, 5 have withdrawn (reasons not recorded) or administratively failed before completion (did not turn in required work), 92 have received a grade permitting them to go on to the advanced seminar course and the remaining 3 have marginally passed. Year enrolments have varied from 5 to 17, for an average of 12. Over the years increasing numbers of students in research-based graduate programmes within geography, or from other departments, have taken the course. In addition, five students have used the course as a formal independent study resource platform, with different assignments.

The course was timetabled in spring 2010 for a fall 2010 initial offering. The author had previously used learning management software (Blackboard© at the time) for information storage and retrieval and assignment submission and return, as well as some required online discussions in large classes. Expanding this experience to completely online took up about one-third of a sabbatical leave in 2009–2010. This time was spent researching the online course support possibilities in general and at the University of Calgary and learning new software applications. Following this preparation time, the production of the course occupied about half time for the 2 months before delivery. Several sessions with institutional-level design and technical support staff helped avoid problems, but no systematic division of responsibilities was contemplated.

The first decision was to use a modified asynchronous delivery. Individual course components could be completed at the student's chosen times and (to some extent) rate. However major assignments including four lab exercises, a term test and a final project would be turned in using common deadlines. The term test would be given synchronously online (see below). These decisions followed online goals and considered the course's place within the larger MGIS programme; they also intended to maximize administrative efficiency within those goals. Improving instructional efficiency was not a goal, although no additional instructional burdens were contemplated, particularly over a multi-offering time period.

Course Format

The course organization, while complex, does not differ greatly from the same course's offline version. Subject matter is divided into "units" collecting three to seven "modules", which correspond roughly to one lecture per module. One unit gathers information about course structure, policies and rationale. It also contains commonly requested information such as grading policy, where to find technical and content help and how to find components of the course. A very important part of this unit, reinforced in course announcements, is how to acquire and install the software necessary to carry out laboratory exercises and the course project.

Each subject-matter unit (e.g. "Specialized Per-Pixel Operations") is introduced by a document placing the unit within the science of remote sensing and situating each module within the unit. Following the first offering, a quiz was added to each unit in multiple-choice format. This quiz is marked automatically and recorded; however it does not count towards the course grade. This allows it to be a purely formative evaluation and to serve the individual student's needs for self-testing and later review both during and after the course. This format allows "wrong" answers to be more ambiguous and to highlight corrections to common misconceptions, which would not be the case with a graded quiz.

The primary structure of the units takes the student from no previous remote sensing knowledge through most commonly used concepts and procedures. A final unit introduces several application areas (such as change detection, radar remote sensing and rock classification) that touch the application interests of large numbers of students in the MGIS programme. Students are only required to complete one module (of their choice) from this final unit. With this structure, students who are brushing up on previous remote sensing knowledge can move quickly through the early units and take on additional application modules. Or, they can devote additional time to elaborating their term project. Similarly, students coming to the discipline for the first time can spend additional time on, or repeatedly return to, earlier modules as required. As the course is offered repeatedly, with new application interests appearing, additional modules can be added to this unit to accommodate these needs, without imposing additional work on all students.

Each unit contains an "extras" section and a "how-to" section. The "extras" adds non-essential information that answers commonly asked theoretical questions; the "how-to" provides background for students who may have missed out on concepts not strictly within remote sensing, such as linear scaling. The how-to modules often directly address the theory behind practical questions raised during lab work or on the discussion board.

Modules do not differ much from lecture notes. Each module contains a "why you should use this module" as well as what concepts are most crucial for understanding and what other modules contribute to understanding the concepts. The module consists of recorded lecture audio, in-module quizzes and some self-demonstration projects with questions attached.

Software

Adobe Presenter v. 7, integrated with *Microsoft PowerPoint* 2002, allowed ordinary PowerPoint slides to be used with the recorded lecture. Video delivery was rejected due to bandwidth limitations, though animations were used where pedagogically advantageous, in the same way as in a classroom. *Presenter* combined with Blackboard© courseware also allowed SCORM content[1] to be integrated. Updating the course for the 2014 delivery used *Microsoft PowerPoint 2013* and *Adobe Presenter v. 10* and was deployed using the D2L (now called Brightspace) learning management system (LMS).

[1] http://scorm.com/

A classroom presentation would contain demonstrations of image viewing and analysis procedures; online the students use their own software to demonstrate these tasks. The self-demonstration results must be used to complete SCORM-based quizzes embedded within the lecture at that strategic point. Quiz completion automatically transfers to the online grade book, allowing the instructor to trace module completion and view individual student responses. Uncertainties about how this would work led to the decision not to include these grades as part of student formal evaluation. To capture similar work for final grades, students are required to discuss questions on the discussion board arising from the results of module quizzes.

With the emphasis on professional software as an integral part of learning, a package was needed that had complete basic and advanced functionality, could be acquired individually by students at low or no cost, worked on most computers students would be expected to have and needed minimal administrative involvement. Online download was optimal. Since internationally based students could potentially take the course, the software chosen would need to be geographically unrestricted. At the time of the original course development, choice fell on *IDRISI Taiga v. 16* (updated in 2012 to *IDRISI Selva v. 17* and in 2014 included in the broader *TerrSet* package). It met all criteria and also supplied tutorials and demonstration data. The university did not have a site license to this software, although individual researchers, including myself, were familiar with it. This meant that students did not have the option of working in an on-campus lab in case of personal installation difficulties. This software did not exist in a Mac-usable version, and attempts to install on PC-emulation Mac segments did not always work out well. Nevertheless, these problems proved to be minor and were student-solved.

Although there were few difficulties with IDRISI products, the lack of local support and licenses in the university labs inconvenienced many students. Also, international portability proved not to be important to the course. Therefore, starting in fall 2014, the course switched to *PCI Geomatica 2014* (and regular updates). This choice among other lab-installed software rested on considerations not directly related to the fact that 603 was online, so it is not further discussed here. Switching software entailed changing specific instructions and some vocabulary used in the modules. This was done together with normal updating.

Some students were already familiar with other software. Where requested, we allowed them to use this provided that they did not insist on adaptation of the module material and vocabulary to that product. This option was ultimately chosen by few students: most who requested it decided after a short while to acquire the common software anyway.

Labs

Since basic software skills were learned through in-module exercises, the laboratory exercises became the prime way to demonstrate and be graded on critical thinking. The goal of labs is to teach students to solve technical and conceptual conundrums at levels that escalate throughout the semester and to teach them to evaluate the correctness and appropriateness of the output. Each lab produces output used in future labs and constitutes a step in small research projects. Thus the students are led into increasing independence by building confidence through the sequence of exercises.

The first laboratory exercise leads students through selecting and downloading a complete Landsat image of their choice, to serve for all subsequent labs and often for the course project, as well as for in-lecture self-demonstrations. Using individual images minimizes plagiarism in labs as well as helping keep student interest. In early labs, specific how-to instructions are included. In later labs, similar tasks are referenced as being similar in structure and sequence. Students are directed to personal experimentation, tutorials and help files to ascertain parameter choice and appropriateness of algorithm selection. In later labs, only completely new sections of the software, or somewhat opaque terminology, are instructed in detail.

Students are encouraged to discuss problems among themselves and to use the online discussion board. This has been successful: students report hardware problems and perplexing error messages and solve them together online. In some classes, one student has become "the help person" and is encouraged to use help requests as a teaching experience ("troubleshoot this way") rather than a tech consulting experience ("push this button").

Each lab requires each student to submit an "informal" presentation-style report about their lab to be made available to everyone in the class. The stated target audience for these is student colleagues with less remote sensing expertise. This online component fills the purpose of work-related lab banter or community conversations. These submissions are not marked directly: non-submission results in a penalty on the lab grade. Pedagogically, the instructor can look at these for hints at shared misunderstandings. The possibility of a synchronous online sharing of these was considered, but the inherent asynchronicity of the course made this unwieldy. Once D2L became the courseware, students could also choose to include these reports in an electronic portfolio. This function had not been conveniently available to Blackboard© users in the university configuration used. More recent courses have provided students the possibility of an on-campus day to share these presentations on a volunteer basis, as a lead-in to general questions and discussions. This has been used by about half the enrolled students and has been appreciated by those who participate.

Integrity, Plagiarism and Time Use: Grades and Statistics

Ideally, skill building can be easily delivered online, serving those who are committed to acquiring it for their own and society's intellectual and economic good. These ideals are easy to achieve using formative evaluation (Flagg 1990), with feedback that reinforces and corrects present performance to improve future performance. Online, this can translate into self-guided quizzes that can be retaken at will, discussion board participation and, in this course's case, the in-module quizzes with automated feedback.

Two items militate against realizing this ideal: the need for objective evaluation of achievement for third-party purposes and for advanced coursework preparation and the need to maintain value and transferability of the course and programme credentials. Resources devoted to the course by the institution (creating, supporting and maintaining it), and course fees paid by the student, need to be validated by measurable output in the form of grades and credits. These considerations favour summative evaluation and may for some students and situations work at cross-purposes to the formative evaluation. The two risk coming into conflict when the summative evaluation becomes part of a student's cost-benefit analysis for time allocation purposes. When many students are also pursuing other demanding courses, research projects and paid work deadlines, this analysis may severely curtail effort invested in an online course that relies primarily on formative evaluation and credits simple completion of retakable quizzes. The 603 summative evaluation includes a large percentage of formal written reports that are individually graded with emphasis on technical writing and individual feedback. This is completely unchanged in the online version as compared with the traditional classroom version.

The asynchronicity of the online course can go some way towards decoupling grades from agenda juggling, since it allows students to proceed at their own pace and arrange their own schedule. Nevertheless, problems remain. Underestimation of the time required for an "automated" task in a lab not uncommonly results in neglect of purely formative evaluation. Writing a formal report may take longer than anticipated as well. In 603 this is tracked through module quizzes completed, course logon times and other course statistics. The statistics available depend on the learning management system used, but data on individual and collective time logged on, as well as time of day, are available in most packages. There is little in a conventional course to give this information. Since 2010, in every year, the beginning of the course sees most students working over lunch hour or early in the evening. Later, times become increasingly diffused (students juggling commitments), and towards due dates many more log on very late at night.[2] There has never been a logon between 5 and 6 a.m. local time.

[2] All statistics are reported in the local time zone of the university server. This course has never enrolled a student more than two time zones away, so these figures fairly represent all students' local time. In a course enrolling students from around the world, LMS statistics would need to convert time zones. To my knowledge no 603 user's group social media page has been established outside of the course structure.

Term Test and Project

The most direct opportunity for cheating in an online course occurs during timed tests. Potentially, the course could enrol students from quite different educational traditions with different understandings of what constitutes "cheating" or plagiarism. Either students must be entirely trusted to understand and voluntarily conform (such as the traditional "honour codes") or systems must be put in place to minimize the possibility of code breaches. This course chooses the latter but attempts to do it as unobtrusively as possible, avoiding proctored exams at distance or synchronous on-campus testing. The one term test is designed around critical use of information, rather than around retrieving facts. Students must perform software functions, interpret results and effectively communicate them within a limited time frame. Students are allowed any resources they want to use, including web access and their course notes and lectures. They are requested to informally cite their outside sources. The one possibility that cannot be tightly controlled would be a student having someone else take the test for them. Since a relatively small percentage of the course grade rests on the test, it was judged that the potential benefit to a student of this kind of cheating would not outweigh the potential cost of discovery. In addition, in a small class with numerous written reports, any notable deviation of the test answer style from the written reports would likely be noticed by the marker. This is a trust that can be afforded in a small course with considerable textual interaction between student and marker; it would not translate into online courses in general, especially large-enrolment ones. The problem online is identical to that of traditionally offered large vs. small courses and fact-based vs. process-based testing objectives. In a process-based small course, cheating is not seen as a particular difficulty. Plagiarism issues in formal written assignments are no different in online than in traditional courses, and precautions and responses depend as much on course size as on presentation format.

Because the course project is independently carried out and results in a graded formal written report, it does not differ substantially from a project in a traditionally delivered course at this level. Disadvantages to online delivery would be the lack of opportunity to talk informally with other students and the instructor before and after lectures. Advantage of offline is that the student has an incentive to discuss the project asynchronously, primarily through email. The discussion boards also allow for voluntary student interaction. Other social media might be leveraged to encourage informal consultation, but this has not been incentivized in 603, and students have not requested it.

Problems and Their Solutions: The Weekly Email

In the traditionally delivered 603, students' main difficulties revolved around being available in person at the right times. The most acute manifestation was in decreased attendance at class sessions and requests for deadline extensions as the semester progressed. The corresponding worry for online was that the complexity of the course, and the difficulty of managing multiple tasks, would result in a high dropout rate or at least in magnifying typical problems of deadline creep. This proved less of a problem than feared. During the first year of operation, the most common email received by the instructor was for reassurance that the student was keeping up. This reassurance could also potentially be accomplished through instructor blog entries or using some other social media. However to keep an already complicated course simpler, the email through the courseware was selected for these communications. The "weekly class email" was born.

A rough schedule of optimal completion rates was developed, based initially on experience of the course's offline offering and later modified with online experience. These recommendations were also placed on the LMS calendar, though there was some fear that they would be perceived there as hard requirements rather than guidelines. The current practice is that each Monday the instructor sends out an email to all students. It lists upcoming hard deadlines and then refers to an ideal completion rate of modules and which specific modules would be most important to the lab work. It also contains harder warnings, for example, "Anyone who has not yet downloaded and activated their software is in great danger of falling irretrievably behind". Then some words of general encouragement or admonition are added as appropriate. The email usually ends with some recent remote sensing related news or perhaps a recent image. On occasion, it might add some interesting ideas (or cautions!) derived from lab marking or the informal lab reports, but it is not intended to be the vehicle for assignment feedback.

These emails prompt group responses and discussions, which are transferred to the online discussion board in many cases. They enabled students to develop and use their own time management skills. The communication of scheduling anxiety decreased markedly after emails were instituted. The emails were least useful for students not following the basic course trajectory, for example, those brushing up on basics or concentrating on more advanced work or those compressing coursework because of time away in inaccessible areas. However, students in these situations did not communicate annoyance at receiving the emails, instead indicating that they understood and appreciated the news and images.

The project was also broken down into submission of a proposal, an outline and bibliography and a completion date, to ensure steady progress. Since these, along with lab due dates, were formal deadlines, they were easily incorporated into the timetable and did not require any specific adaptation to online offering.

Staff Time and Effort

In addition to serving students' needs and freeing up lecture room space, online courses can be thought to decrease overall costs per student without sacrificing – in fact while enhancing – learning quality. Potential for success of this strategy seems more obvious in very large lecture classes, where the instructor's lectures can be encapsulated and delivered efficiently, while in-person instructional time can be used to leverage personal or small-group interactions responding to that content. Are there other increases both in efficiency and effectiveness that could apply to a high-interaction, small, advanced-level class, where traditionally the student-instructor interaction has been highly valued? The experiment with this class was also undertaken to answer this question, particularly in the case where the subject matter itself is in a fact-laden field. Can an online course produce critical interaction with technical "facts" while being a better use of the instructor's time and expertise than lectures and demonstrations? Or does it just make more work for the instructor with no efficiency or cost benefits to them or to the administrative apparatus?

It is theoretically possible to present an online course using a combination of existing independent services such as YouTube, a blog hosting service, Twitter, email and the like. This was rejected as being a return to the early and inefficient form of web use in teaching in the 1990s. In practical terms, to even contemplate an online course, the sponsoring institution must have a LMS in place, with automated template creation, student enrolment linkages to the registration process, security provisions, content management, a document submission system and a grading system. How these are organized, and what additional services are needed, will vary from institution to institution and from course to course. Most courseware includes some form of grading and feedback mechanism, file management and scheduling tools, plus internal communications and discussion possibilities.

The institution must agree to schedule a course as online. Where registration is integrated with university-wide classroom assignment and student timetabling utilities, this may present more problems than anticipated if the institution is not actively fostering conversion to online courses. Finally, the institution must have some sort of support to answer LMS and hardware questions for both students and instructors. Most institutions will offer LMS training and customization and instructional design and support. These last however are not always present and, unless there is a large local push for online delivery, may be limited to a few experimental systems and personnel that collectively do not greatly diminish the time commitment of the primary instructor. Any specialized software outside of the LMS would require additional time and support from the instructor or instructional team, and this guided our choice of image processing software.

This course, being small and at an advanced level, did not benefit from online tools for either intensive discussions or large class handling. The lengthy 1-year (part-time) period taken to develop 603 was very useful for finding and avoiding known pitfalls. Once the framework had been established and tools acquired and tested, the actual course building took place over about 3 months, again not full time. The majority of this time was taken up in recording lectures, integrating the

SCORM content and testing all systems with the cooperation of a few research students. For the first offering, an extra about 2 h per week, beyond the time devoted to delivering the course, was devoted to troubleshooting, minor revisions, evaluating on the fly and planning more major revisions. An informal survey to gauge student reaction was conducted after the second offering was finished. This was not envisaged at the time as publishable research, rather as course improvement, so no ethics approval was sought.

In the first year, the time usually devoted in a conventional class to immediate class preparation and delivery was balanced by online course upkeep tasks such as developing the weekly emails, monitoring and contributing to discussion boards and checking statistics. Marking took neither more nor less time than for an offline course. Thus the first delivery following course development was time equivalent. Subsequent offerings benefitted from lessons learned, and overall upkeep tasks were less time consuming. The time devoted to subject-matter revision was about the same as for a subsequent conventional offering. This yielded time commitment proportional to the number of students enrolled, meaning a much shorter time for a smaller class. In an environment with fluctuating student numbers, and a demand to keep numbers up to justify instructional time, this is a benefit: the saved time can be devoted to other courses or duties without penalizing the online students. This is a surprising result, where a system touted as making large classes more efficient ended up in being able to justify a smaller class! The instructor, like the students, benefitted from asynchronicity insofar as it allowed easier schedule control and diminished multitasking.

Additional efficiencies from online delivery became apparent only in the second year. Since the modules cover basic and intermediate remote sensing topics in self-contained units, they can easily be turned into "learning objects" (LORs) usable in other courses, at least by the same instructor. This is discussed further below. Also, a university department often has undergraduate and graduate students who request special consideration for various valid reasons to take required courses outside of the scheduled time period. Small combinations of course modules can be extracted and presented to these students to replace the lecture portion of an undergraduate remote sensing introduction course. The "extras" and "how-to" 603 modules can be formally added to an undergraduate course. Added to undergraduate-level labs and tests, a single student can be accommodated with no more additional effort than the marking required, which can be assigned to the undergraduate course teaching assistant. Denial of requests for such a course usually occurs because of the instructional and administrative time involved in an individually tailored course. The cannibalism of online course elements as LORs eliminates many objections. This option however is only approved for near-graduating students with an acceptable GPA, given the self-regulation requirements of working online in a complex course.

Another use of the materials is for graduate student review and general reference. Students report referring back to the modules when encountering practical problems in their later courses or research or when preparing for a comprehensive examination. The course materials can also be made available to programme students wishing to audit or to self-teach, with no additional costs to the administrative system. The only problem to be solved to benefit from these uses is to maintain access of students to the LMS for an extended time period after the end of the course. Students are not able to access the course after leaving the university, however. This would be a matter of university policies, not related to the course itself.

Student Evaluation

This course exists primarily to prepare students for an advanced graduate seminar in remote sensing. This seminar's objectives are to explore various application areas and become familiar with cutting-edge research topics. After the first offering of 603, the seminar instructor made some suggestions about content adjustment, which were carried out. It was useful for other instructors to refer to the modules to be able to see exactly what their students had used previously. No systematic differences between online 603 students and other students were observed, although no formal investigation was carried out. There did not seem to be any grade difference in the early years between those without 603 and those with it, though small numbers precluded any statistical analysis. Later seminar offerings had too few non-online 603 graduates to reach any meaningful conclusion. The 603 instructor also taught the advanced seminar twice following the implementation of the online version, in 2013 and 2015, and observed similar good preparation, particularly in the independence and critical abilities of students.

An exit survey was conducted for course improvement purposes after the first three offerings. This was not intended to constitute publishable research and was not covered by ethics approvals. Therefore no quantitative results are reported. Students' preferences for online or offline had to do with their personal preferences for face-to-face vs. online interactions in other areas of their lives. Some preferred the more social atmosphere of a classroom, and some preferred the computer screen. It is to be noted that this distinction primarily occurs between students who say they prefer small classes vs. those that prefer large lecture sections. This reinforces the direction of online education as replacing large classes. Nevertheless, no students stated that they felt penalized by the online delivery, and many added that although they preferred the in-person small class, they did recognize the advantages of online delivery, primarily asynchronicity.

Formal instructor evaluations were problematic, in that the questions (uniform throughout the campus for all courses at all levels) were largely oriented towards in-class teacher presentation evaluation. Since the results for this class would not be compatible with those for an in-person class, and since the numbers were small, it was decided not to administer these evaluation questionnaires to students in the online class. If online classes were to occupy a much larger portion of the timetable, a separate questionnaire would need to be developed. The course provided anonymous routes of commentary or complaint should a student want them: these have never been used.

The existing courseware registration and grade reporting systems were not modified. It was inconvenient that "TBA" was entered on the university timetable, leaving students in the dark about on-campus presence requirements. Some students thought they had a choice between the same course in-class and online. These concerns led to increased questions during registration, to the instructor, departmental administration and student advising services. For this small course, a note to the people concerned solved the immediate difficulty. This was later supplemented by a timetable footnote saying the course was online only. For a larger number of courses, or for larger enrolment courses, this informal solution would need to be replaced by a uniform system of notation explaining the logistics to students as part of their registration materials.

Students came to know about 603 through the MGIS programme requirements. Some students outside the programme in closely allied departments (Archaeology, Engineering, Ecology, Biological Sciences, Geology), who wanted an intensive introduction to remote sensing, also enrolled starting in the first year of offering. These external students usually heard about the course from fellow students or from their academic advisors. Occasionally someone in a job where working knowledge of remote sensing would be useful has registered for the programme. This relied on word of mouth. Communications campaigns were not deemed worthwhile for a single course.

Where 603 Online Fits into Pedagogical Trends

Experiential learning Online learning has a fraught relationship with "experiential learning". Kolb (2015) defines experiential learning as "learning where the learner is in direct touch with the realities being studied" (p. xviii). Remote sensing is the topic of 603. By definition, it uses information acquired by sensors *not* in direct contact with the object, hence contradicting the "direct touch". However this course makes use of the experience of the student in acquiring satellite imagery from data centres and using self-selected entire images to carry out self-selected procedures to arrive at a coherent analysis. This is in fact the "experience" of the operational remote sensing professional, and so is what students need to be in direct touch with. Remote sensing of some applications does require direct gathering of field data for validating its results. This course therefore can be seen as experiential in an indirect sense, or it may confer skills that are important adjuncts to the field skills appropriate to an application field such as forestry or agriculture.

Blended learning Blended learning is a term used very broadly to include many combinations of technology with non-technology-based styles (Graham 2006). Technology is not necessarily the same as online, but as more procedures migrate to cloud-based computing, they are converging. The blended learning paradigm recognizes that individual instructors may choose online and/or technological components that best enhance their teaching practice. With the adoption of centralized learning management systems in most universities in Canada, all courses are potentially blended. Attention turns to the toolsets used rather than the concept of blended learning. The 603 online course was not intended to be blended, as no component of it required students to be physically present. After the first 603 offering, however, students were given the option to ask a question of the instructor in person rather than by email or discussion board, if the student was located on campus. As will be seen below, later offerings moved towards provision of an optional in-person lab session. The course started online and became blended!

Flipped classroom A flipped classroom is one where traditional "lecture" components, where desired, might be pre-recorded, simulcast or archived so students need not be present (Kim et al. 2014). The times for meeting in person would be devoted to direct problem solving, group work or highly interactive sessions. Because 603 was developed requiring students to access the course online, the concept of a flipped classroom was not initially included in the planning. Nevertheless it has been gradually moving towards a flipped model for laboratories and project development. Because of the original requirement for students to be able to take the course without being on campus at all, the flipped components have remained optional and not assigned specific grades. The advantages of in-person meetings are made available to students able to participate, once per month. For example, students prepare short "popular" presentations reflecting on their learning motivations and experiences, which are submitted with graded assignments and made available to other students. They may use any technology they wish for these, from basic PowerPoints to whatever media is available within the LMS. These can be presented and discussed by students physically present and seen online by students not available at that time period. Also along the flipped classroom continuum, within the recorded lectures, students create computer-based "case studies" that take the place of in-lecture instructor demonstrations of how things are done. Student participation is verified using immediate feedback and scoring related to questions they answer about the results of their experiment. This introduces a long-distance component similar to the active experimentation that flipped classrooms carry out during the face-to-face sessions.

What is a "course"? In 2016, discussion is ongoing about the relevance of a "course" as the basic unit of learning. This was well expressed in an online comment:

> The idea of a course, I think, has two great categories. The first is a kind of intermediate simultaneity – we don't always have to focus on the subject on Mondays, Wednesdays, and Fridays from 11:15 to 12:05, but we can only spend time on it between January 8 and April 29. The second is its formality – that is, everything that derives from having a defined structure, including assigning instructors, bandwidth, and other institutional resources; … and granting widely recognizable forms of credit.
>
> Learning doesn't have to take the form of a course. If we really wanted to promote learning, we'd never use course forms at all (although there might be some value to structuring experiences so that groups would start together and promote peer-to-peer learning). But the course is the way that we recognize learning within the formal collegiate

> structure. So a learning experience must accommodate the formalities of a course if it is to be a part of collegiate learning[3].

The offline 603 was redesigned for online presentation in 2010 intending for it to remain a standard "course"; it was neither a pilot for an online programme nor a vehicle for intended further restructuring of many courses. All of the considerations in the above quote apply to it: asynchronous at the daily time scale but synchronous at the semester scale and fitting in to existing credit, registration, grading and administrative systems. The additional advantage of traditional course structure is that it can be managed by a single course instructor. It could both enhance learning and solve other organizational problems without massive support from outside the course structure itself.

Massification The 603 course is small and advanced and so not a prime candidate for engaging in economies of scale or "massification". Course outcomes focus on the student's ability to use and adapt tools to meet theoretical and practical objectives. However 603 is also technical in nature, requiring that students learn many computer-based skills, some of which can be massified. The skills taught in 603 are not necessarily software-specific; most mass online instruction in remote sensing is software-specific in the form of tutorials, webinars and manuals. Many online general "tutorials" have existed in remote sensing since the 1990s. These may be more like traditional textbooks (e.g. Short's (n.d.) tutorial[45] and the Canada Centre for Remote Sensing (n.d.)[6]). Software-specific tutorials exist that concentrate on the basic functions of the image processing package. Many software companies offer online webinars highlighting improvements and additions to their functionality. Many of these are aimed at workplace and research lab upgrading of skills and so may be closely woven with promotional material to the neglect of theoretical understanding.

Online programmes Many universities and colleges now have entire degree or certificate programmes of study offered online. These integrate most course components into a system including the courses and centrally organized marketing, administration, course design and advising systems (for one example among many, the Penn State World Campus[7]). While these programmes are in some sense "competition" for the MGIS programme of which 603 is a part, the individual courses within them were not direct models for 603, which was converted to online for internal purposes.

---

[3] Comment by "sibyledu" self-identified as A. Michael Berman, in response to J. Kim, "DOCS not MOOCS" on *Inside Higher Education* online blogs: Technology and Learning, February 3, 2015. https://www.insidehighered.com/blogs/technology-and-learning/docs-not-moocs

[4] All URLs as listed were accessible in early 2016.

[5] http://www.fas.org/irp/imint/docs/rst/Front/overview.html

[6] http://www.nrcan.gc.ca/earth-sciences/geomatics/satellite-imagery-air-photos/satellite-imagery-products/educational-resources/9309

[7] http://www.worldcampus.psu.edu/RS

Future Directions: Towards a Flipped Classroom

Starting in 2014, 603 reserved a 3-h/week block in an on-campus computer laboratory. The required software was installed, and technical support was at hand should it be needed. Students were not required to be present at these sessions, to maintain the possibility of asynchronicity. Their purpose was as a possible backup if students' software installations or hardware crashed. Regular instructor walkabouts at these times revealed usually 3 or 4 students (out of 16) making use of the lab, usually the same people each time. This system costs nothing (others may use the labs at the same time), so it has been maintained.

With few long-distance registrations to date, the course recommends itself for a modified "flipped classroom" approach. The on-site portions would not require attendance, again maintaining optional total asynchronicity, even for students in local work situations. The class offering in fall 2016 and 2017 included some additional incentive for attendance at one or more lab periods.

Online scheduled office hours using multiple connections and tools such as Skype are sometimes used with flipped courses. However such a system would further limit asynchronicity of the course and would meet no obvious need. All these potential flips will be managed to maintain the course's ability to serve working and travelling students while not diminishing services to those who remain on campus.

Future Directions: The Concept of a Small Online Course

The preparation, operation and pedagogical experience of this course have been presented and also informally discussed in conference settings (Hall-Beyer 2012). Most of the questions and conversations in that setting concerned administrative support and technical difficulties and possibilities, rather than the teaching and learning experience itself. One might conclude that the largest barrier to adoption of an online approach to the small advanced class remains at the institutional level, in providing impetus, support, systems adaptation, time for preparation and advertising for such courses. This may change in future, as online education is a rapidly morphing field. The model used does not, however, scale to a MOOC or other large-enrolment model. It is not likely to be of great interest where budget restraint or teaching efficiency is a large motive for going online. On the other hand, the model adopted does not incur any expenses beyond the traditional course and if used more widely would eventually cut down on local space and hardware use, allowing some savings or diversion of resources. To balance this, more technological and design support would be helpful, especially for first-time course instructors. The model's main advantage would be increasing student and faculty flexibility and therefore satisfaction, without costing a great deal. The primary investment required has been the learning curve in planning and organizing the course. This diminishes for subsequent online offerings by the same instructor or as online capabilities become more generally known by the professoriate.

Making an online version of 603 has been an experiment. The course itself can be considered a success, because it has continued to provide the required level of expertise to its students. Some instructional efficiencies have been realized and crossover benefits accrued; the course has been successfully migrated across courseware platforms, software upgrades and analytical software changes. All of this has been accomplished in no more time than "ordinary" course planning and delivery. Course organization and upkeep have remained the purview almost exclusively of the original instructor. While other instructors, and the larger university teaching community, have evinced interest through workshops and conversations, no increase in online courses offerings within the programme has occurred. The instructor has become a resource for others more in the planning and effectiveness as opposed to the technical process of course creation.

The course was offered online for 2016 and 2017. Later, as it transitions to other instructors, or as the underlying programme evolves, its continuation as an online course will depend on the interest and willingness of other instructors to update and develop it. As the developing instructor, I would be interested in adapting other courses of similar format to online delivery, providing that practical and policy support is forthcoming. It works for students. On the other hand, this course has not demonstrated any overwhelming advantage that would strongly recommend online use for economic or policy reasons. Entirely online programmes exist and have the backup and reach required to successfully compete for students seeking out such a programme. A single course within a larger programme will not likely attract students specifically interested in online study.

Future Directions: Learning Objects

Learning objects (LOs) refer to development of pieces of material that are self-contained enough to be used within contexts other than those in which they were created. Another common term is open educational resources (OERs). According to Wills and Pegler (2016), the concept is developing rapidly to include design elements and other aspects of reusable materials. When seen in this way as an efficient way of reusing material, LOs are positioned within larger context of multi-instructor courses, multi-university and multi-jurisdiction collaboration in teaching and learning and for-profit courses and programmes. At this broader level, questions surround not so much the content of a module or unit as defined for 603, as they raise issues of format standardization, evergreening, access and ability to edit and above all intellectual property. We are not concerned here with these wider issues but with facilitating reuse and repurposing of the existing modules and units and indeed the ability to repurpose the course as a whole for particular delivery needs.

LOs can be defined many ways, but one commercial manager of them defines them as "adaptive, competency-based learning technologies[8]". This puts the emphasis on the process rather than the content. At a different level of detail, the British Columbia Ministry of Education defines them as "self-contained educational resources that are properly tagged with keywords, or other metadata, and often stored in an XML file format[9]".

Each module in 603 is structured to start with a "why you need this module", a review of topics and concepts necessary before starting and a note on how it relates to other topics within the same unit. This suggests that the 603 unit would be the proper scale to be transformed into a stand-alone LO, with the addition of exercises, self-quizzes, a smaller lab and report and a graded outcome. Each unit could quickly become a building block for other courses. They could be put together in various combinations and have minor alterations for particular needs without compromising the pedagogical structure.

Currently, LO repositories (LORs) exist in some LMS, which limits their use to that system[10]. Independent repositories, as stand-alone companies and attached to educational media, exist[11]. Some public jurisdictions are developing LORs although the format and legalities have not yet settled down (Richards et al. 2002; McGreal et al. 2004, see also footnote 9). Many of these start with K-12 level LOs to accord with jurisdiction-wide curricula. So far LORs are not heavily invested in graduate-level materials, likely because of the small-group or individual coaching model applied to most graduate courses. A paucity of scholarly work since an initial interest in the early 2000s indicates the state of flux of LOR status. The main roadblocks are software common to all potential users, practical control of access, maintenance of integrity of material (and attribution) while allowing adaptation, updating procedures, use in fee-bearing credentialing (who gets the fees and who controls the quality), intellectual property rights and control and similar issues.

Aside from receiving grants to develop LOs for curricula mandated across institutions, it is not yet obvious what advantages exist for the individual instructor to contribute to or to use LOs. Until LORs are stabilized, it is to individual instructors' advantage to produce LOs on a small scale, initially for that instructor's own use in current and future courses. Limited sharing among department or allied department members, all using the same institutional LMS, would be easy to arrange and would not pose the larger organizational and intellectual property questions. Later, these LOs might be easily adapted to a university-wide catalogue, announced on a research or teaching social media site or maintained by governmental or professional bodies. There are currently the usual disincentives attached to early small-scale adoption of

[8] http://www.learningobjects.com/

[9] https://www.bced.gov.bc.ca/dist_learning/learning_objects.htm

[10] http://www.brightspace.com/products/learning-repository/ is attached to the D2L courseware used in 603.

[11] http://www.editlib.org/noaccess/20615/ is one example in a rapidly shifting world. No endorsement is implied and no attempt is made to survey the current offerings.

a rapidly evolving system. Nevertheless, the experience with the 603 course indicates that a personal or local LOR can have its advantages and pre-position the instructor to take advantage of larger-scale LORs as they become available.

An informal LO system has allowed 603 or portions of it to be made available to students either as an individual-study course or as individual modules and units called upon in response to a student's detailed need arising from research or preparation for another course. This has allowed administrative flexibility with very little additional instructional time. To be successfully deployed in this way, the students involved need to assure themselves and the instructor that they can function independently without the scheduling and peer group framing that occurs during the regular semester course offering. This kind of offering eliminates the advantages of the flipped classroom aspects of 603 but allows the material to be effectively taught and evaluated in unusual situations.

Summary

The 603 course was developed for students at a graduate level but contained introductory level material pertaining to the science and technology surrounding remote sensing data, theory and methods. It replaced a face-to-face course. The changed format helped students with scheduling, especially for part-time and travelling students. It did not in fact fulfill the possibility of having international students take one course in their home country after enrolling in the larger programme of which the course is a part. This is not a structural problem with the online course but rather that this demand did not materialize. At the discipline level, the course produced students who performed as expected in an advanced seminar course. Popularity, as judged using informal surveys, proved high for some students and average for others; the rate of withdrawal or failure was no different from the traditional offering and was very low as is typical for graduate-level courses. There was no enduring consistent pattern of likes and dislikes by students, and their constructive comments were easily incorporated into minor changes in the course format. The main difficulties overcome related to students feeling isolated, especially with regard to expected progress and software learning. This was overcome with additional framing of the timetable and active encouragement of online and email communication both with the instructor and among students. An optional on-campus lab period was initiated but was used by only a few students. This confirms the utility of an asynchronous format.

An additional unforeseen advantage was the ability to adapt the course to custom-design special needs courses without much added instructor time. This encourages further development of the course to produce smaller self-contained "learning objects" made out of modules and assignments, able to be incorporated into other courses by the original instructor or by others. Full deployment of such LOs will require solution to administrative and technical challenges relating to open or semiopen learning object repositories.

Future directions for the course include optional synchronous events, moving in the direction of a partial "flipped-classroom" course while maintaining the desired asynchronicity. The programme in which the course is situated is not expected to develop into a completely online format because of requirements for design and management support, as well as extensive marketing, beyond that available to individual instructors. Overall, the successes have been notable though limited in scope and can be recommended as an effective way to increase instructor efficiency without penalizing – in fact while benefitting – advanced students.

References

Canada Centre for Remote Sensing (n.d). https://www.nrcan.gc.ca/sites/www.nrcan.gc.ca/files/earthsciences/pdf/resource/tutor/fundam/pdf/fundamentals_e.pdf

Chen, B., & Denoyelles, A. (2013). Exploring students' mobile learning practices in higher education. *Educase Review.* http://er.educause.edu/articles/2013/10/-exploring-students-mobile-learning-practices-in-higher-education

Flagg, B. N. (1990, reprinted 2013). *Formative evaluation for educational technologies.* New York/London: Routledge.

Graham, C. R. (2006). Blended learning systems: Definition, current trends and future directions. In C. J. Bonk & C. R. Graham (Eds.), *Handbook of blended learning: Global perspectives, local designs.* San Francisco: Pfeiffer Publishing.

Hall-Beyer, M. (2012). Online teaching of remote sensing: A technological nightmare or a technological fit*? Canadian Association of Geographers annual meeting* 2012, Waterloo, ON, May 28–June 2 2012. Abstract at www.cagacg.ca/files/pdf/agm/2012_AGM_program.pdf

Kim, M. K., Kim, S. M., Khera, O., & Getman, J. (2014). The experience of three flipped classrooms in an urban university: An exploration of design principles. *The Internet and Higher Education, 22*, 37–50.

Kolb, D. A. (2015). *Experiential learning: Experience as the source of learning and development* (2nd ed.). Upper Saddle River: Pearson Education.

Liyanagunawardena, T. R., Adams, A. A., & Williams, S. A. (2013). MOOCS: A systematic study of the published literature 2008–2012. *The International Review of Research in Open and Distributed Learning, 14*(3), 202–227.

McGreal, R., Anderson, T., Bain, G., Downes, S., Friesen, N., Harrigan, K., Hatala, M., MacLeod, D., Mattson, M., Paquette, G., Richards, G., Roberts, T., & Schafer, S. (2004). EduSource: Canada's learning object repository network International. *Journal of Instructional Technology and Distance Learning, 1*(3), http://www.itdl.org/Journal/Mar_04/article01.htm

Moore, A. S., & Simon, S. (2015). *Globally networked teaching in the humanities.* New York/London: Routledge. 239 pp.

Richards, G., McGreal, R., Hatala, M., & Friesen, N. (2002). The evolution of learn object repository technologies: Portals for on-line objects for learning. *Journal of Distance Education, 17*(3), 67–79.

Short's (n.d). http://geoinfo.amu.edu.pl/wpk/rst/rst/Front/overview.html

Tess, P. A. (2013). The role of social media in higher education classes (real and virtual): A literature review. *Computers in Human Behaviour, 29*(5), A60–A68.

Wills, S., & Pegler, C. (2016). A deeper understanding of reuse: Learning designs, activities, resources and their context. *Journal of Interactive Media in Education, 2016*(1), 10, https://doi.org/10.5334/jime.405

Web GIS in Development: From Research and Teaching Perspectives

Ruibo Han

Abstract With the prevalence of the Internet and mobile devices, the visualization and presentation of geographical data are not limited to paper-based maps anymore. Geographical Information Science (GIS) web services are the software components that host spatial data and GIS functionalities that can be accessed and integrated into customized GIS applications through the Internet. Developers utilize GIS web services for custom applications that process geographical information without having to maintain a full GIS system or the associated spatial data. Two key benefits of web-based GIS distribution systems are the increased interaction with users and connections to a wider audience and its advanced data integration capabilities. With a number of projects and cases in Canada, the potential of Web GIS is demonstrated from a research perspective in the fields of migration, communication, culture, etc. The fast development of Web GIS will not only help improve the research in academia but also has the potential to benefit public society as a whole. Therefore, the transmission of capabilities in Web GIS to students via university teaching is also discussed. Web GIS has brought up inexorably changes to the teaching of GIS. These changes present both opportunities and challenges for educators and students.

Keywords Data visualization · GIS server · Mobile GIS · PPGIS · Web GIS

Introduction

Geographic Information Systems

Geography, generally considered to be one of the world's primitive scientific disciplines, has always embraced new skills and technologies and been practiced in research and instructional areas.

R. Han (✉)
Department of Geographical Sciences, University of Maryland, College Park, MD, USA
e-mail: ruibo@umd.edu

S. Balram, J. Boxall (eds.), *GIScience Teaching and Learning Perspectives*,
Advances in Geographic Information Science,
https://doi.org/10.1007/978-3-030-06058-9_7

Canada has been a pioneer in developing geographical information systems (GIS). Since Dr. Roger Tomlinson's Canada Geographic Information System in the 1960s, GIS has been widely used as indispensable tool to collect, store, manage, analyze, and display geographic data. GIS can be used to build the inventory of various types of data with spatial attributes, which have seen an expanding usage in both human topics (e.g., demographic databases) and natural studies (e.g., distribution of environmental elements and factors). GIS also provides a growing and large number of tools that can be used to analyze and assess the characteristics of data over data space, temporal space, and spatial space. Another impressive characteristic of GIS is its usage in visualizing and presenting the result of spatial phenomena and analysis on maps, even in three-dimensional view.

The challenge for geography teaching in the twenty-first century is how to teach the subject effectively while sticking tightly to its traditional heritage and embracing emerging technology and fast-developing tools and perspectives. The education of GIS transmits a theoretical and practical knowledge for careers in the field. GIS degree and certificate programs and technical courses are available at nearly every major university in Canada, at many technical colleges, and countless of online programs.

As stated by Tomlinson (2007), "three legs of the stool of future development" are "technology, people and data." This not only lists the importance of the three factors in future development but also illustrates the connections among factors: the accessibility of technology to the general public through education system, the accessibility of data to the public through data, as well as people's understanding and sharing about data. In fact, the general public is becoming more spatially exposed to the usage of digital maps for directions, travel, and exploration on computers, tablets, cell phones, etc. There is an increasing awareness of the interconnection of people and the environment, coupled with increasing understanding that the pervasive problems of our world need a holistic and geographic approach to solving them, where Web GIS has the advantage.

The Potential of Web GIS

With the prevalence of the Internet and mobile devices, the visualization and presentation of geographical data are not limited to paper-based maps anymore. Much recent attention has focused on developing GIS functionality over the Internet, Worldwide Web, and on the cloud, and this branch of GIS technologies deployed with Internet is termed as Web GIS. The terms Internet GIS and Web Mapping are also commonly used to describe GIS capabilities on the web to complement or even replace the analytical capacity of desktop GIS.

Web mapping applications were initially developed to publish static maps over the Internet, but more sophisticated technologies and applications soon joined the family with the development of Internet technology and hardware. Xerox was generally considered to be the first to publish maps via the Internet in 1993 and soon followed by the Tiger Mapping Service in 1995, which allowed users to graphically query census data published by the United States Census Bureau (Manson et al. 2014). The number and variety of Web GIS application emerged rapidly as technology improved thereafter. Google maps together with Google Earth released in 2005 stand out as a prominent example of Web GIS applications.

Based on the power of GIS, Web GIS holds the potential to distribute geographic information to a larger worldwide audience. Internet users will be able to access GIS applications from their browsers without proprietary GIS knowledge or GIS software. In addition, Web GIS makes it possible to add GIS functionality to a wide range of network-based applications in business, government, and education. For instance, geospatial data and analysis tools are being distributed over the Internet, and interactive cartography is available to the general public to display customized data or results.

Teaching with Web GIS allows for innovative and authentic technologies, tasks and assessments, and also improved problem-based, team-based learning. Incorporating Internet and mobile GIS tools can help students reach the top of the pyramid in Bloom's taxonomy (Bloom and Krathwohl 1984) (Fig. 1):

The teaching of Web GIS at the college level has been grouped into two categories: (1) application-based learning and (2) programming-based development. The first group focuses on the lower levels of Bloom's taxonomy, including teaching the concepts and theories about Web GIS and introducing various technologies for mapping, disseminating, and analyzing GIS data via the Internet or in the cloud. A large number of tools and platforms are employed to achieve this, and both server-side platforms (e.g., ArcGIS server, MapServer, GeoServer) and client-side tools (e.g., ArcGIS Online, AppStudio, Web AppBuilder, QGIS) have been extensively

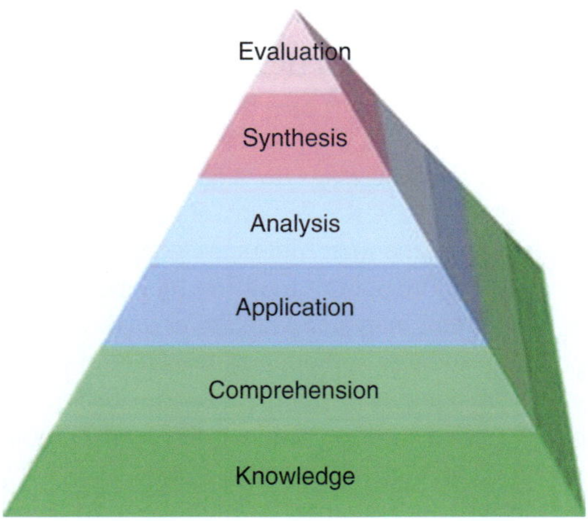

Fig. 1 Bloom's taxonomy of six levels within the cognitive domain. (Source: https://sites.google.com/site/prakashbebington/instructional-design-resources/blooms-taxonomy)

applied in case studies or class projects. The second group focuses on the development of new tools and interfaces for Web GIS analysis and mapping. The higher level of coding skills are required for such courses, and various languages and platforms (e.g., ArcGIS API for JavaScript, Flex and Silverlight, CartoDB, and Leaflet) are also available to create a versatile Web GIS development environment. The products from these groups are generally equipped with a customized interface and higher-end analysis tools. Though commercial software, such as ESRI products, are generally the major tools to be taught in college curriculum due to the stable performance and superior product support, open-source Web GIS tools are gradually being introduced to the classroom. Open-source GIS has been set up as an undergraduate/graduate course in many GIS programs.

Web GIS Advantages

Robust and Versatile Deployment Strategies

Many strategies can be employed to implement GIS functionality to the Web: Server-side strategies allow multiple users to submit requests for data and analysis to a GIS Web server. The server collects and processes the requests and returns data or results to the remote user. Client-side strategies allow the users to perform simple data management and analysis locally on their own machines, such as editing a table or uploading ancillary information. To enhance the performance of Web GIS platforms and meet special user needs, hybrid strategies that combine server and client ones are commonly deployed.

Web GIS developers can design and code their applications from scratch based on project needs and technologies available. Another commonly used method is purchasing the necessary GIS modules from commercial vendors and directly populating project data on those established Web GIS systems.

Open-mindedness and High Extensibility and Flexibility

Web GIS makes it possible to share data and technology in an open-minded environment. Commercial GIS and Web GIS solutions typically require a systematic investment on a series of software (ArcGIS, ArcGIS Server, ArcSDE, etc.), and the maintenance and upgrading may generate additional costs. However, Web GIS systems have more options to utilize the open-source data and Web GIS platforms, such as open street map and MapServer as development platform, and adopt open technology standards and support the OGC protocol, which supports a variety of GIS data format and distributing platforms (Mingyi et al. 2013). In addition, Web GIS is also featured by high extensibility and flexibility of platform architecture of the system. The web has become a data and thoughts sharing platform that features bottom-up information uploading with increasingly spontaneous spatial information.

Web GIS Applications

Applications that utilized Web GIS as major or minor research tools have flourished over the past decade, and these applications are valuable resources in the classroom. They have been studied as reading materials and exercises for understanding Web GIS platforms, and they are also classic prototypes to be followed for class projects.

Data Visualization and Presentation

A major contribution of Web GIS is the visualization and presentation of results. A chart is worth a thousand words, but in geographical studies a map is worth a thousand charts, and in Web GIS these charts are broadcasted via Internet channels. As an important component of GIS, cartographic mapping is capable of adding more explanatory and intuitive power to the traditional tabular and graphical reporting formats. The importance of this capability in ecosystem assessment and management is reflected by the persistent reference in the adaptive management literature to the significant communicative role of clear visual presentation of results (Holling 1978; Environmental and Social Systems Analysts Ltd 1982; Walters 1986).

GIS Web services are the software components that host spatial data and GIS functionalities that can be accessed and integrated into customized GIS applications through the Internet. Developers utilize GIS Web services for custom applications that process geographical information without having to maintain a full GIS system or the associated spatial data (Li et al. 2011). Users can tap into Web-based GIS distribution systems through their Web browsers without having any specialized GIS software on the desktop system. Web-based GIS technologies have also enabled the possibility of using the Internet to publish the data from the inventory and GIS database. The focus in this mode of GIS use is not necessary on its geodetic or analytic capabilities (although they do play a major role) but rather on the visual and contextual exploration of the problem situation and issues connected to it.

Case Studies: Mapping Migration from the Americas

The Mapping Migration from the Americas project provides information about the spatial distribution of foreign workers in Canada from Latin America and the Caribbean (LAC) region allowing users to analyze the relationship between migration flows and development impacts. Mapping Migration from the Americas was launched in 2008 by the Canadian Foundation for the Americas (FOCAL) and includes interactive maps of Canada and also regional maps of Colombia, El Salvador, Guatemala, Jamaica, Mexico, and Trinidad and Tobago.

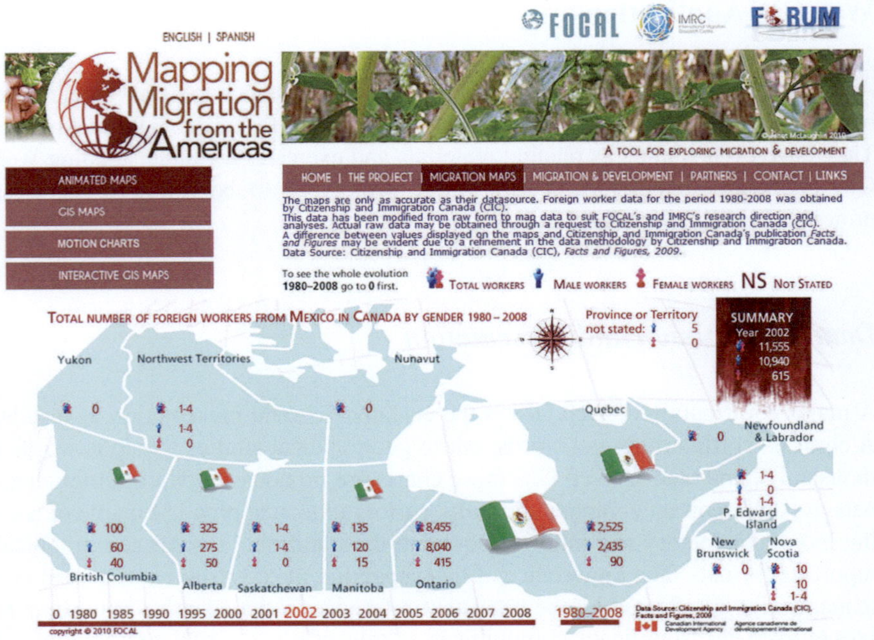

Fig. 2 Animated maps are predesigned flash maps that can demonstrate the transition patterns of the number and distribution of foreign workers over years

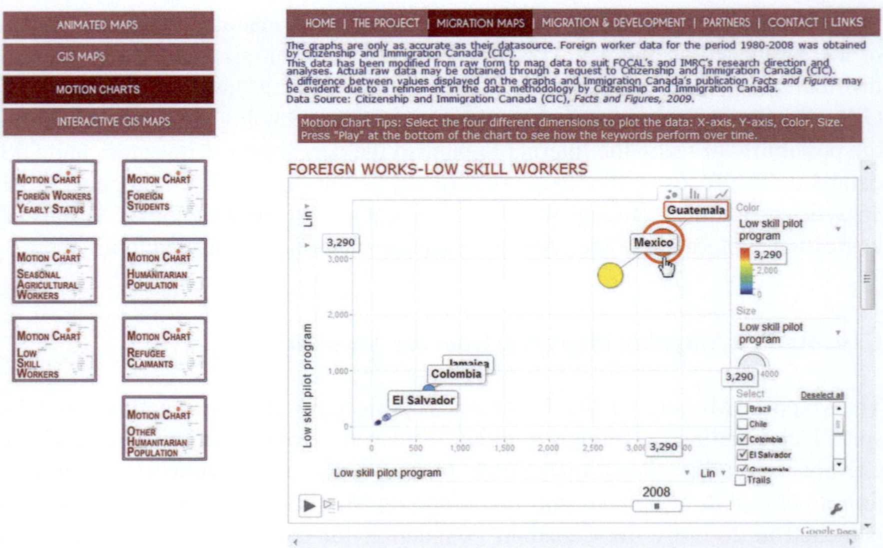

Fig. 3 Motion chart is an interactive tool that compares the number of foreign workers from Latin America and the Caribbean region over years

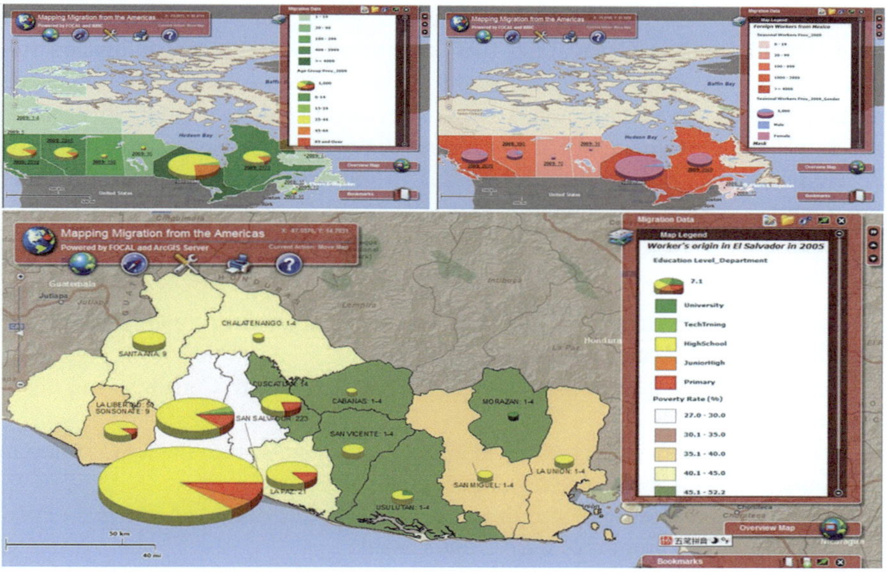

Fig. 4 Interactive mapping interface

The Mapping Migration initiative uses Web GIS technology to bring together previously unconnected databases of migration information that was for the most part unavailable to the public.

This is the first time that data on the temporary workforce coming to Canada from Latin America and the Caribbean region is displayed along with its spatial distribution across Canada and in sending countries. Migration and other data can be projected onto the same map, providing context while allowing users to visualize possible relationships and patterns within the datasets. The maps feature a variety of tools and functions that can help support research and analysis (Figs. 2, 3, and 4).

In the interactive mapping interface, datasets can be layered and crossed for an at-a-glance illustration of the connections between the distribution of temporary foreign workers in Canada and information such as country of origin, sex, age, education, and occupation, where available. Users can also explore the maps by zooming in for detail or by panning across maps to visualize geographical differences. For in-depth analysis, users can perform refined searches and sort results using the query function.

The Web GIS portal is able to print customized maps that can be used in publications, presentations, and educational materials, if properly referenced. The website also features a section with country reports, links to information pertaining to migration programs, and additional information on the different thematic areas linked to migration and development, such as remittances and health.

The maps can be used by a variety of users in their efforts to better grasp the relationship between migration and development. The combination of geographical, statistical, and quantitative information can help policy-makers and migration experts develop new policy options to address issues related to temporary labor migration that impact Canada and LAC countries. Likewise, academics and students can use the maps to conduct studies on the influence of migration flows on development in the LAC region. Civil society organizations can use this tool to access valuable information that can inform dialogue initiatives on migration and development and work with policy-makers to improve temporary foreign worker programs.

The maps will increase the capacity of experts and researchers in the LAC region and Canada to:

- Analyze the movements of temporary workers and uncover trends between the LAC region and Canada
- Identify important emerging issues for the LAC and Canadian migration agendas
- Deepen the understanding of circular labor movements from the LAC region to Canada and their associated development impacts
- Assist policy-makers in identifying alternatives for the design of effective migration policies and programs in LAC and Canada

Connection Between People and Data

Web GIS is also capable of building the connection between people and data. Two other key benefits of Web-based GIS distribution systems are the increased interaction with users and connections to a wider audience (Kyem and Saku 2009), and its advanced data integration capabilities (Kulawiak et al. 2010). Thus, there is potential for more people over a broader domain to be reached through the Internet than other forum options and certainly at a lower cost compared to traditional methods – i.e., printing or public forums. In addition, updates to data can be made on the Web server and are immediately available to users with little or no printing costs.

Case Study: Mapping the Media in the Americas

The Mapping the Media in the Americas project investigates the relationship between media and democracy in Canada and 11 Latin American countries. The project provides information about the location, coverage, and ownership structure of the media (television, radio, and print media), along with key information about demographics and election results. Mapping the Media in the Americas was launched in 2004 and includes maps of Argentina, Brazil, Canada, Chile, Colombia, Costa Rica, the Dominican Republic, Guatemala, Mexico, Peru, Trinidad and Tobago, and Uruguay.

The media plays an important role in the decisions that citizens make in democratic societies. Information about elections, candidates, public services, and government policies is transmitted through the media. Mapping the Media in the Americas highlights the link that exists between the media, the citizens, and their electoral choices, helping to understand how they influence each other in Latin America.

Creating an environment that promotes transparency and freedom of the press is impossible without analyzing these linkages. Despite this need, however, there continues to be an absence of accurate information about the media in many countries throughout Latin America. Little is publicly known about the ownership structure of the media, the impact of media messaging on the vote or the effect of media concentration and its potential threat to the diversity of ideas, freedom of expression, and access to information. The media maps have been created in order to foster transparency and enhance knowledge about the role and connection of media with democracy and provide access to valuable data and resources for students and researchers in order to further investigate these important connections. The project aims to map the location, coverage, and ownership structure of the media (TV, radio, cable, and print media) in 12 countries in the hemisphere, crossing this data with electoral results and sociodemographic information. The maps bring an innovative application of Web GIS to the social sciences and employ cutting-edge interactive software to convert normally static maps into Web-accessible, interactive maps. The maps bring together previously unconnected, and often not public, databases of media, electoral, and sociodemographic information to make them available for use and reference by the public. While these data are just a start and must continue to be improved and expanded, they represent an important advance for transparency and access to information in the Americas. The Mapping Media project is always in motion – as we work to develop more maps and integrate new and interesting data from around the globe, beyond America. With this in mind, and in an effort to advance toward the overarching and long-term goal, the project has set aims to:

1. Strengthen the technical capacity of Latin American partners to sustain and develop the maps in keeping with domestic needs in the long term
2. Support the integration of the maps in fact-based research, analysis, and practical applications
3. Promote the maps among a wide range of stakeholders as a resource to further explore the realities of media and its relationship to democracy in Latin America (Figs. 5 and 6).

Integration of Spatial and Nonspatial Datasets

Another key benefit of Web-based GIS is its capability of distributing systems to relate a wide range of spatial and nonspatial datasets. These systems have been used as public forums or as decision support tools for various projects from

Fig. 5 Mapping the Media in the Americas

environmental assessments to transportation infrastructure and mass transit routing (Li et al. 2011). These systems can integrate spatially referenced shapefiles with tabular attribute data, satellite imagery, and aerial photographs. In addition, other photographs, images, and documents along with links to additional Web resources can be incorporated. Common analysis tools also allow users to extract, overlay, and join spatially related data, create buffers and service areas around features, and perform advanced spatial and network analyses.

Case Study: The Promised Land Project

The Promised Land Project (PLP) is a million dollars 5-year (2007–2012) Community-University Research Alliance (CURA) focusing on "historical amnesia" vis-à-vis the contribution of nineteenth-century black pioneers in Chatham, Chatham Townships and Dawn settlements, and the role this multicultural group of

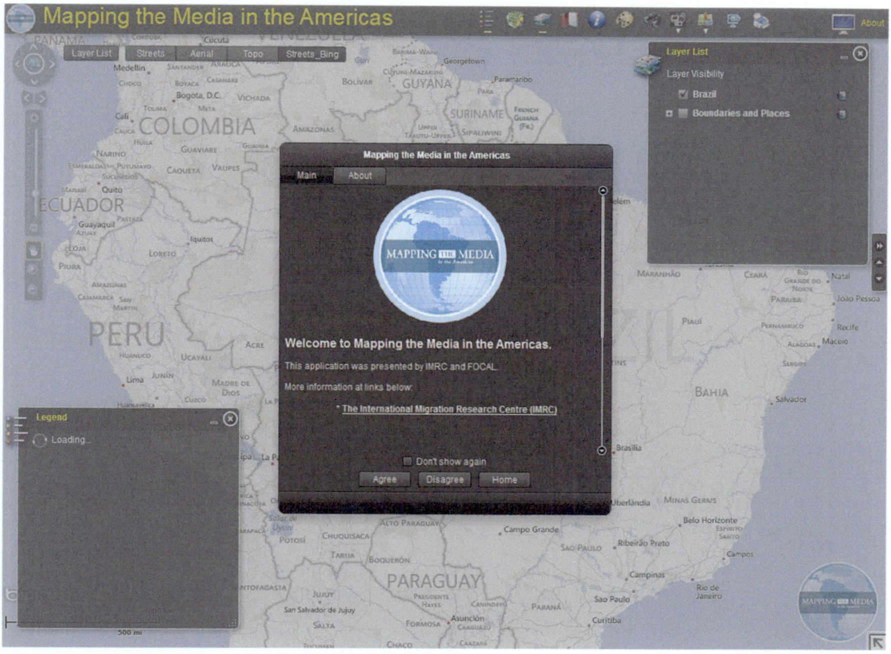

Fig. 6 Interactive Web GIS interface

blacks, whites, and Natives played to end slavery and to fight for civil rights in Canada, the United States, and abroad.

Grand narratives of Canadian history refer to the Promised Land localities, merely as the final stop on the Underground Railroad. This narrow "escape from slavery" narrative does not explore what happened once so-called fugitives had found their freedom nor the experience of blacks in the Promised Land communities. Though the communities themselves were small, their experience and influence stretched across Canada and to the farthest reaches of the Atlantic world. It is this vital center of culture and justice that drew interracial support and forged trans-geographical links of freedom between Canada, the United States, and United Kingdom that this research project would like to integrate into the Canadian grand narratives.

This Web-based GIS portal is a successful application of Web GIS in social media studies. This website allows the researcher to publish media files including a comprehensive database of letters, tax records, journals, photographs, oral histories, family narratives, newspapers, and other important primary sources. The users cannot only interactively communicate with the maps by examining and querying in the current database, they are also allowed to upload maps and media documents. The newly added social media (Youtube, Flicker, Twitter) tool is also of great interest to those who are social media followers.

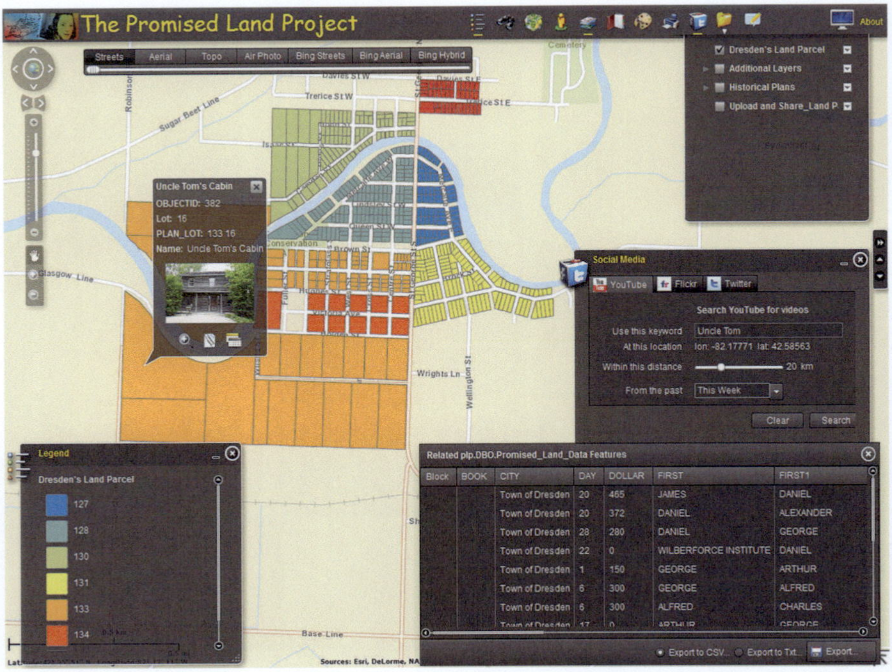

Fig. 7 Web GIS application in The Promised Land Project

This Web-based GIS allows users to search the repository through selection criteria with a series of menus and query functions to retrieve data results for assessment or management. The output is displayed through a combination of text, maps, and digital orthophotos. This system is thus a comprehensive data delivery tool that can assist policy-makers and stakeholders with development decisions to encourage sustainable management of the region (Fig. 7).

Public Participation GIS (PPGIS)

Web-based GIS also facilitate the procedure of collecting inputs from the public in the process of ecosystem assessment and management. The integration of user requirement and user feedbacks is now indispensable in general information systems design (Onwuegbuzie et al. 2010) and in GIS design (Brown 2012). Public participation GIS (PPGIS) has emerged as a test bed for techniques, methodologies, ideas, and discussion about the social implication of GIS technology. PPGIS enables users to benefit from GIS' ability to bring together many different data sources into comprehensive and manageable format, which is an excellent tool for data management. For instance, community groups and citizens can contribute information such as historic land uses, old photographs, or other data that completes the inventory of an ecosystem.

Case Studies: Mapping Spatial Injustice

The issue of urban sustainability has been increasingly acknowledged as an issue of importance, particularly the role of social actors as agents for sustainability and the importance of social cohesion. The city of Ürümqi, capital of Xinjiang Uyghur Autonomous Region of China, has been the site of conflict and social tension between the marginalized indigenous Uyghur minority and the Han majority group for decades. Under rapid urbanization, Uyghur and Han ethnic groups have come into more intensive contact in the city and have developed spatially segregated ethnic enclaves. Social tension has recently manifested in several incidents of open conflict. Using this pertinent case study of a socially unsustainable arrangement, which can be applied to many similar cases in the world, including Canada, this research has the following objectives:

- To identify the spatial factors involved in building resilience in an ethnically diverse urban setting and to identify vulnerabilities relating to the exclusion and inclusion of different ethnic groups
- To understand the role of agency, power, and knowledge in achieving social resilience and sustainability
- To adapt participatory methodological approaches for use in the comparative study of spatial perceptions, to be available for both academic and nonacademic settings

One method to include local spatial knowledge, particularly the knowledge of those at a power disadvantage, is the *public participation geographic information systems* (PPGIS) approach. PPGIS uses geographic information systems to incorporate knowledge of underrepresented communities along with other spatial and nonspatial data. The element of spatial knowledge inclusion from marginalized communities offered by PPGIS relates directly to some of the key factors in resilience theory and also to understanding the production of urban space. PPGIS is used to conduct a comparative analysis of Uyghur and Han perceptions in neighborhoods of Ürümqi. The project includes 4 months of fieldwork and data collection in Ürümqi, as well as data analysis and writing in Canada. Six neighborhoods in Ürümqi featuring different ethnic mixed are selected for participation.

The study is based on a quantitative and a qualitative methodology, in an attempt to measure and compare the spatial perceptions of inclusion, exclusion, and conflict as experienced by Uyghur and Han people in their lived experience in Ürümqi. This spatial analysis gives insight into social and spatial factors that create vulnerability to the crisis among the city's population. It primarily involves a double level of analysis: **an area-level analysis** whose purpose is to define a macro perspective on the variation of social space from one neighborhood to another, particularly the neighborhoods that have been the site of conflict and social tension. It allows us to identify patterns in terms of socioeconomic inequalities across Uyghur and Han areas and some of the potential structural factors underlying these disparities. However, an area-level analysis is limited in its explanatory potential to the extent that it overlooks the local specificities and circumstances that may influence individual perceptions of *spatial injustice*. Moving away from a reliance upon government documents, and in order to capture these specificities, this research relies upon **an individual-level analysis** at the neighborhood level, whose purpose is to better understand the challenges that local minority (and majority) populations face in terms of their lived experience in their neighborhoods. In order to obtain a micro-level image of spatial injustice, a Web-based PPGIS platform is provided to residents of Ürümqi from our six sampling neighborhoods in order to record their individual perceptions of spaces in which they feel excluded and included and places where they have seen/experienced conflict.

In addition to the macro level analysis, a Web-based PPGIS platform is designed. A simple, interactive, online platform is created in which users can manipulate city/neighborhood maps of Ürümqi to create spatially located data. The website is in three languages (English/Uyghur/Mandarin and French) with instructional videos for guidance. In collaboration with XNU, a total of six neighborhoods are selected, consisting of two group samples each consisting of three neighborhoods with very different socioeconomic status between both groups. The two groups include a set of neighborhoods that are Uyghur majority, Han majority, and mixed Uyghur-Han, in order to differentiate the perceptions between neighborhoods that are of different ethnic makeup – representing the enclave nature of Urumqi's sociodemographic layout. The two groups vary spatially, with the first group of neighborhoods being situated in the city center in close proximity and the second group of neighborhoods scattered in more in the peripheral areas of the city. Furthermore, 25 residents in each of the 6 neighborhoods are selected as a core focus group. This core group of residents, along with representatives from neighborhood residents' committees, local government, and the city planning bureau, attend a series of community workshops. Issues of social tension, urban development, and the PPGIS platform are discussed. In particular, it is asked of Uyghur and Han people to define what it means to be Uyghur/Han living in Xinjiang, how locals think cultural identity is manifest, and under what circumstances (if any) their ethnic identity is not engaged. This is particularly addressed from a cultural standpoint to better understand what each group views as important in creating space that is culturally relevant. The PPGIS will undergo troubleshooting based on interactions and suggestions.

Integration of Other Fields of Geomatics into Web GIS

As subdisciplines of geomatics, both GIS and remote sensing are technologies that focus exclusively on geographic data to represent the world's geographic features. Technically and conceptually, each technology is established on diverging principles and applications, where remote sensing is principally used as a technology to collect data, while GIS is one that is dedicated to handling, processing, analyzing, and presenting data. Therefore, the integration of remote sensing into Web GIS will produce a representation of the world as reliably and realistically as possible and as wildly and timely as a possible atmosphere.

Remote sensing is an ancillary source of information for GIS databases. In Web GIS, remote sensing data has been widely used as base maps and references for GIS data display, which only takes a minimal advantage of the power of remote sensing. There is still great potential that remote sensing can be more directly used in Web GIS analysis with the development of the Web GIS technology itself. For example, GIS and remote sensing together can implement advanced analytical functions. These include attribute- and spatial-based queries, the overlay analysis of statistical and thematic attributes from both GIS and remote sensing, Boolean and fuzzy logic analysis, and the building of complicated models using multiple analysis tools (Mesev 2007). Wang et al. (2015) designed a map-based, spatially correlated design method for a computer-assisted instruction system in geography. By integrating spatial information technology (geographic information system, global positioning system, remote sensing, etc.) and information delivery methods (Web services, databases, etc.), a Web GIS-based teaching assistant system for geography field practice was established, realizing an effective spatial management scheme and forming a shared platform for instruction material.

The public's growing desire for mobility and convenience has driven the propagation of mobile Web devices, led by the smartphone and tablets, which is rapidly becoming a strategic platform as a Web GIS client. Mobile devices will surpass desktops and notebooks as the primary client platform for Web GIS, and location-based services will be provided not only for both corporations and consumers but also for governments, academics, and related stakeholders. A cross-platform Web-based development framework for mobile GIS and remote sensing applications was introduced by Tsou et al. (2005) for laptops, tablets, and mobile phone platforms. Using these platforms, Java software technology was examined for its cross-platform utility in the development of various mobile GIS functions.

Despite the technical advances in geomatics and the expanding applications in research, there are still a number of fundamental issues that remain unaddressed when using remote sensing techniques in Web GIS. For example, there is little theoretical consideration of finding the most appropriate spatial resolution (scale) for remote data while considering the domain of the study area. In addition, Web GIS still lacks many statistical functions. Users often have to use external statistical desktop software in order to run some statistical analysis.

Opportunities and Challenges of Web GIS in GIS Education

Web mapping is widely used by a large number of parties in various fields, and interest continues to grow in employing it in the classroom, but many questions remain about the resource demands of this technology and its value for GIS education (Manson et al. 2014). The emergence and recent flourish of Internet-based GIS is going to, and has already done so in some universities, slowly but inexorably change the teaching with GIS. These changes present both opportunities and challenges for educators and students.

Opportunities and Benefit of Web GIS

Internet-based GIS provides numerous advantages that may deliver the ability of educators to meet spatial learning challenges (Baker 2005). A much larger volume of geospatial tools, which require less training to master compared to desktop GIS, are available for teachers. These tools are mostly deployed using a Web browser, and this serves a significant benefit in terms of the saving of investment on hardware and software to construct traditional GIS systems. Additionally, the simpler interface and operations on Web GIS require less time to learn, thus dramatically reduce the consumption of time and money for educators to incorporate them into curricula. Furthermore, the increasing awareness of open data policies greatly reduces the restrictions of sharing data and contributes to the accessibility of data and knowledge over the globe.

Web-based GIS tools can be effectively used to enhance teaching and learning experience in combination with traditional GIS and remote sensing. Numerous tools from desktop GIS and multiple sources of remote sensing data can be incorporated into Web GIS platforms to mash up a comprehensive GIS portal. Future Web applications will deliver a much more expressive, intuitive, fun, and fast-responding user experience, due to the advances in computing power, Web browser technology, and GIS plug-in technology. This has attracted a surge of interest from both professional and student GIS users to adopt the available Web resources and build mash-up applications.

Web GIS can be used in a variety of ways to solve real-world problems using updated data and methods. Web GIS makes it easier and more practical to deploy an authentic GIS system in other disciplines, and educators are including Web GIS into multidisciplinary curricula, such as environmental studies, biology, sociology, economics, and history, to name a few. Web GIS can be applied at any spatial and temporal scale – from analyzing locally collected data about individuals at the local scale to aggregating social problems about a group of entities at mesoscale and to studying global issues and problems at a macroscale. Web GIS is useful anytime the "where" and "when" questions are asked, and the answer to these questions can be explored or displayed over the Internet. In each application in education, students can learn about the processes to model the patterns of location and place on our planet and investigate the reasons why these patterns exist and processes occur.

Challenges to Web-Based GIS in Education

Opportunities in Web GIS are also coupled with challenges of learning and teaching them. The challenges exist in both using and developing Web GIS in education systems.

Challenges in using Web GIS in education involve uncontrolled quality of data, limited functionality of Web GIS, strict Internet security rules, the requirement of Internet bandwidth, pedagogical shift for teachers, and distractions for students using Web tools in class. First, the easy procedure of publishing GIS data promotes the sharing of data but also creates the problem of uncontrolled quality of GIS data online. The lack of proper training on abiding metadata standards and a consistent censorship of published data contribute to the wealth of GIS data of mixing quality, which may post challenges for users to determine the value of data. Second, Web GIS is still under constant development, and most of the current applications are equipped with functions of visualizing data but lack the functions of analyzing data, though Web GIS is not developed to substitute desktop GIS due to the limitations of hardware and computing power. A related disadvantage for Web GIS use is the requirement of fast Internet bandwidth and customized securities rules. As the data and maps are all stored on the cloud and streamed to clients upon request, the user experience and functionality rely largely on a fast and stable Internet connection. Most applications require the installation of plug-ins (java, flash, Silverlight, etc) for a smoother and visually appealing user interface, and these plug-ins might conflict with university's rules of security or lack of permission of installing software on lab computers. Lastly, education in Web GIS may involve pedagogical shift for instructors who are using to traditional GIS educational system and may require further training to adapt the challenges for interpreting and analyzing Web GIS data. For students, the benefits of Web GIS in class might be weakened if they are captivated by the appearance of Web tools or Web access instead of using the tools to solve problems.

Beside the existing challenges using Web GIS in class, there are also major ones if advanced courses are introduced to design and implement Web GIS platforms. Deploying Web GIS requires a higher standard of hardware and Internet bandwidth and also customized server machines with server software and packages. This system also requires constant maintenance and technical support while students are learning to deploy Web GIS. In courses of mobile GIS developments, the development package (Android Studio, Xcode, etc.) requires higher-end PC or Mac machines with more user permission, and handheld devices are additional investments for educators as well. Furthermore, higher programing skills are the requirement to teach Web GIS classes. Though Web GIS programming environment is filled with easy-to-use development tools, a good knowledge of multiple programing languages is a prerequisite for teaching advanced Web GIS classes. Commercial Web mapping solutions are provided from companies such as Intergraph and Environmental Systems Research Institute (ESRI). Open-source Web GIS platforms such as MapServer and GeoServer also emerged in the mid-1990s, followed later by free Web mapping offered by private enterprise sites such as Google Maps and ESRI's ArcGIS Online. Browser-side GIS APIs, such as ArcGIS API for JavaScript, ArcGIS API for Flex, and ArcGIS API for Silverlight, have greatly simplified Web GIS application development but still require programing skills to manipulate with data and maps. The visual design of the Web GIS interface also requires care and aesthetic design to assure that users can understand and make use of the information and functions provided by the system. Lastly, "learning is the process whereby knowledge is created through the transformation of experience" (Healey and Jenkins 2000). The programing tools and platforms receive constant upgrades and updates with the development of computer science, which is good for the Web GIS technology in general, but would require instructors to spend more time on refilling knowledge tank.

The abovementioned challenges prevail among most Canadian universities, colleges, or private training institutions, and most of these challenges are not easily solved with software and hardware. In fact, most universities have yet provided advance Web GIS or mobile GIS courses due to these challenges, which has adversely affected the development of the GIS discipline.

Future of Web GIS in Education

The development of GIS in recent decades has been revolutionized by rapid development of hardware and software; however, Web GIS in Canadian education still needs advocacy. While Web GIS technology has made significant advances during the past decade, university curricula for GIS are unable to keep pace with these developments in general. There is a fundamental need to update undergraduate and graduate GIS programs to reflect the prominence of Web GIS and promote Web GIS within the university curricula.

Future paradigms to incorporate Web GIS into current curricula can be categorized as Web GIS education and Web GIS-involved education (Fu and Sun 2010). Web GIS education focuses on Web GIS technologies, principles, and applications. Mobile GIS based on handheld devices is considered as a subfield of Web GIS education. Due to the increasing value and applicability of Web GIS in areas of science, governments, business, etc., practical knowledge and skill of Web GIS provide students with a competitive advantage in the future job market and also benefit or even determine the future of employers.

Web GIS-involved education focuses on applying ready Web GIS platforms in solving a spatial problem in multiple subjects. This is also a cost-effective way to expose GIS technology to other disciplines and expand GIS education in general. There is a growing amount of free data, map services, analysis tools, and applications available over the web in various fields of study. The audience of this education can range from elementary through graduate school or even professionals in needs of this technology. With intuitive Web maps, geo-spatial and geo-temporal analytic tools, and diverse platforms and adaptability, Web GIS-involved education can advocate spatial thinking and help students solve real-world problems.

Aside from the traditional classroom education in the GIS technology arena, there has been a fast emergence of online GIS/Web GIS education via webinars and open online courses that are free of charge, as well as training workshops and certificate programs with charged services. The attendance of online courses generally produces a document/certificate with credentials. To add more entries to the formal degree system, accredited certificates are gaining popularity. These certificates have lower requirements and are more focused on practical and recent skill-sets and typically cost less in terms of time and money than a traditional degree program. Another advantage of a certificate program is that asynchronous learning can easily be implemented to fit the schedule of professionals, who can freely watch recorded lectures and participate in discussion or group activities at their convenience.

Last but not least, the educational use of Web GIS/GIS outside the universities and institutes of higher learning has also been limited. The proper implementation of Web GIS instruction may promote geographic competence and interdisciplinary learning in the classroom, as well as foster a resource-rich environment, enhance spatial reasoning, and support problem solving for younger minds. Despite its potentials, many secondary schools still lack the resources and know-how required to integrate Web GIS into curricula. As discussed by Sui (1995), GIS education involves two aspects, namely, how to teach about GIS and how to teach with GIS. Both aspects require to be reinforced outside of higher education systems, especially in secondary schools.

References

Baker, T. R. (2005). Internet-based GIS mapping in support of K-12 education. *The Professional Geographer, 57*(1), 44–50.

Bloom, B. S., & Krathwohl, D. R. (1984). *Taxonomy of educational objectives book 1: Cognitive domain*. New York: Addison Wesley Publishing Company.

Brown, G. (2012). An empirical evaluation of the spatial accuracy of public participation GIS (PPGIS) data. *Applied Geography, 34*, 289–294.

Environmental and Social Systems Analysts Ltd. (1982). *Review and Evaluation of Adaptive Environmental Assessment and Management*. Vancouver: Environment Canada.

Fu, P., & Sun, J. (2010). *Web GIS: Principles and applications*. Redlands, CA: ESRI Press.

Healey, M., & Jenkins, A. (2000). Kolb's experiential learning theory and its application in geography in higher education. *Journal of Geography, 99*(5), 185–195.

Holling, C. S. (1978). *Adaptive environmental assessment and management*. London: John Wiley & Sons.

Kulawiak, M., et al. (2010). Interactive visualization of marine pollution monitoring and forecasting data via a Web-based GIS. *Computers & Geosciences, 36*(8), 1069–1080.

Kyem, P. A. K., & Saku, J. C. (2009). Web-based GIS and the future of participatory GIS applications within local and indigenous communities. *The Electronic Journal of Information Systems in Developing Countries, 38*(7), 1–16.

Li, S., et al. (2011). *Advances in web-based GIS, mapping services and applications*. Boca Raton: CRC Press.

Manson, S., et al. (2014). Resource needs and pedagogical value of web mapping for spatial thinking. *Journal of Geography, 113*(3), 107–117.

Mesev, V. (2007). *Integration of GIS and remote sensing*. West Sussex, UK: John Wiley & Sons Ltd.

Mingyi, D., et al. (2013). *Educational geographic information system based on WebGIS*. Information science and cloud computing companion (ISCC-C), 2013 international conference on.

Onwuegbuzie, A. J., et al. (2010). Innovative data collection strategies in qualitative research. *The Qualitative Report, 15*(3), 696–726.

Sui, D. Z. (1995). A pedagogic framework to link GIS to the intellectual core of geography. *Journal of Geography, 94*(6), 578–591.

Tomlinson, R. F. (2007). *Thinking about GIS: geographic information system planning for managers*. Redlands, CA: ESRI.

Tsou, M.-H., et al. (2005). A web-based java framework for cross-platform mobile GIS and remote sensing applications. *GIScience & Remote Sensing, 42*(4), 333–357.

Walters, C. (1986). *Adaptive management of renewable resources*. New York: Macmillan Publishing Company.

Wang, J., et al. (2015). A WebGIS-based teaching assistant system for geography field practice (TASGFP). *British Journal of Educational Technology, 47*, 279–293.

Neogeography: Rethinking Participatory Mapping and Place-Based Learning in the Age of the Geoweb

Jon Corbett and Gabrielle Legault

Abstract The term neogeography describes a new approach within GIScience teaching and research whereby geospatial technologies are becoming web-based and increasingly accessible to a broad range of developers and users. Through engaging with the geoweb (geospatial web-based applications), users are creating maps and sharing spatial data that is relevant to their own lives and experiences. The development of geoweb tools has been especially important for students and faculty members who engage in the practice of participatory mapping. In this chapter we describe how, in a university taught neogeography course, undergraduate students created participatory geoweb applications to address local issues. One student constructed a geoweb-based platform for civic dialogue; a second team of students designed and deployed an application to address the logistical challenges faced by a local grassroots organization. We draw on the pedagogic theory of place-based education (PBE) and consider it as an effective approach to teaching and learning neogeography. We observed how a PBE approach empowered students to critically and effectively engage in civic matters, become more invested in 'place' and actively learn directly about social justice issues. The approach also supported an increased academic engagement while providing a valuable learning experience in terms of developing technical, critical thinking and other soft skills.

Keywords Neogeography · Geoweb · Place-based education · Participatory mapping

J. Corbett (✉) · G. Legault
Community, Culture and Global Studies, UBC Okanagan, Kelowna, BC, Canada
e-mail: jon.corbett@ubc.ca

S. Balram, J. Boxall (eds.), *GIScience Teaching and Learning Perspectives*,
Advances in Geographic Information Science,
https://doi.org/10.1007/978-3-030-06058-9_8

Introduction

The term neogeography is used to describe an emerging field in which 'complex techniques of cartography and GIS [are put] within reach of users and developers' (Turner 2006). These changes are demonstrated by the increasing number of geospatial web-based applications (referred to as geoweb), frameworks and resources that together make it easy to create maps and share the locations of one's own interests and history (e.g. Google Maps, OpenStreetMaps, Foursquare, etc.). Through a series of readings, lectures, seminars and a project, we taught a course that critically explores the evolution of neogeography and the geoweb, its increasing prevalence and importance in society as well as its ability to influence the manner in which we consider our relationship to space, place and the world around us. The course is built on seminal texts from the discipline of geography, GIScience, sociology and Applied Internet Research. Central to the course is a term project that requires students to practice problem-solving using neogeography tools. In the cases described, students engaged with local community members and organizations to create a geoweb-based tool to address specific civic challenges. By applying a place-based educational framework, students felt empowered to affect change in the places that mattered to them while promoting social justice and developing both technical (hard) and soft skills. This chapter will report on the construction and delivery of the course and provide two examples of student place-based projects.

Participatory Mapping

Counter to the 'premise that cartographers and map-makers engage in an unquestionably "scientific" or "objective" form of knowledge creation' (Harley 1989) that represents an undeniable 'truth' (Coulson 1977; Dorling 1998), academics and practitioners alike increasingly understand that representation of geographic information is not neutral and is in no way separate from the power relations of society (Livingstone 1992; Kwan 2002). Since the inception of map-making, ruling powers and elites have used maps as a medium to exert their privilege (Wood 1992), as well as 'fix' spatial information around subjective knowledge claims (Elwood and Leszczynski 2011) that may, or may not, present competing and discordant versions of spatial reality (Warf and Sui 2010).

Despite forces that have served to exclude non-experts from map-making, the past 25 years have witnessed an explosion of participatory mapping initiatives throughout the world, in both developing and developed countries (see Poole 1995; Parker 2006; Corbett 2009). Participatory mapping is, in its broadest sense, the creation of maps by members of the public – often with the involvement of supporting organizations including governments (at various levels), non-governmental organizations (NGOs), universities, developers and other actors engaged in land-related issues. Participatory maps provide a process of creation and visual representation of the spatial knowledge that a community perceives as their place and identify features of significance within it.

The practice of participatory mapping often focuses on providing the skills and expertise for community members to create these maps themselves, while the content of these maps represents the spatial knowledge of community members (Kienberger 2014). The participatory mapping process can influence the internal dynamic of a community, contribute to building a community vision, help stimulate community members to engage in place-related challenges, raise awareness about place-related issues and ultimately contribute to empowering local communities and their members (Corbett 2009; Panek 2014).

The general aims and specific objectives of participatory mapping initiatives depend on the end use of maps, as well as the audience that will view and make decisions related to map content. Participatory mapping projects can take on an advocacy role and actively seek recognition for community spaces through identifying land uses, local experiences and environmental considerations. In this form participatory maps play an important role in helping diverse interests in the community such as those held by farmers, indigenous peoples, the homeless, recreation groups, youth and/or seniors to work towards recognition of where their interests overlap.

Participatory mapping uses a range of methods as well as cartographic tools (Brown and Raymond 2014) that include mental maps, ephemeral mapping, sketch mapping, transect mapping and participatory three-dimensional modelling. More recently participatory mapping initiatives have begun to use digital applications most broadly referred to as geographic information technologies (GITs) but including specific technologies such as Global Positioning Systems (GPS), remote sensed imagery and geographic information systems (GIS).

In their current state, GITs remained widely used by government, business, large NGOs and academia but less so by local communities. However, since 1995, critical GIScience research has begun to question this prevalent GIS paradigm and its associated power discrepancies, as well as to explore the potential to deploy and modify the technologies so that they might be used more effectively by communities and other grassroots groups (Sieber 2007). In contrast to more technocratic driven uses of GIS, Elwood et al. (2011) urge critical geographers to place a greater focus on 'doing GIS in critical contexts, extending calls for critical cartographic literacy' (Elwood et al. 2011, 91). This debate coalesces around the academic discourse of public participatory GIS (PPGIS) (Sieber 2006). PPGIS, as Laituri (2003) notes, is 'the confluence of social activity (participatory activities, grassroots organizations, governmental decision-making, the Internet) and technology (computers, hardware, software, digital information, the Internet) in specific places – grounded geographies' (25). Given this definition, it is clear to see that PPGIS, and more broadly critical GIS, shares much in common with the principles of participatory mapping.

Many early examples of PPGIS implementation were located at North America and involved universities working in conjunction with community organizations to use spatial technologies to support local development initiatives (Craig and Elwood 1998; Elwood 2002) using a variety of models to support technology access and provision (Leitner et al. 2002). In some cases these interventions involved using community GIS facilities, but more often they drew on university labs and desktop computer-based GIS as opposed to web-based applications (Elwood and Ghose 2001). Among other outcomes, the results of these projects were used to lobby governments for policy and service provision (Ghose 2007), as well as to identify local assets like parks, community gardens and public libraries. Often these projects involved graduate and undergraduate students and as such represented not just an important initiative for the community but also a meaningful learning opportunity for those students involved.

Neogeography

Macintyre (2008) wrote in the Times of London newspaper 'A new golden age of cartography has suddenly dawned, everywhere. We can all be map-makers now, navigating across a landscape of ideas that the cartographers of the past could never have imagined'. The statement recognizes the emergence of geospatial web-based technologies. In an academic environment, this development is associated with terms such as neogeography (Turner 2006), Web 2.0 (Keen 2007; Sui and Holt 2008) and the geospatial web (Scharl and Tochtermann 2007). From a practical perspective, the statement draws on a growing body of available interactive web-based mapping technologies such as Google Maps, OpenStreetMaps and Foursquare. From the perspective of participatory mapping, the geoweb presents a new potential to bridge the gap between GITs and a truly participatory means of sharing and managing community digital spatial knowledge (Tulloch 2008; Hall et al. 2010; McCall et al. 2014).

Development of the potential for the geoweb to support participatory mapping initiatives has been closely associated with the unprecedented growth in both hardware (in particular mobile technologies) and software capabilities, as well as the burgeoning use of the Internet to provide maps and other geographic information (Haklay et al. 2008). New web programming languages have allowed simple and attractive interfaces to deliver both intricate functionality and complex information. Furthermore, the interactive capability of the geoweb enables new web-based applications to both collect and publish user-generated digital content. Concepts used to describe this user interactivity include terms such as crowdsourcing and volunteered geographic information (VGI) (Goodchild 2007). Scholars have begun to study the potential for the geoweb to support a two-way flow of information between citizens and governing agencies, referring to this process and these tools as a participatory geoweb (Johnson and Sieber 2011).

Previous barriers such as financial cost and the need for specialized training are significantly reduced in the geoweb model, and overall the distance between the roles of contributor, producer and consumer of geographic content has decreased (Haklay et al. 2008). The geoweb offers a new participatory mapping approach for the communication and discussion of spatially bounded community priorities and local knowledge (Rambaldi et al. 2006; Tulloch 2008; Sui and DeLyser 2012). However, caution remains around the potential for the geoweb to be a truly effective and accessible tool for public engagement (Elwood 2008; Jankowski 2011; Kingston 2011) especially in regard to the way that social barriers continue to act as barriers to access (Crutcher and Zook 2009; Stephens 2013).

Undoubtedly, the implications that neogeography and geoweb-based technologies could have for participatory mapping projects will depend upon how projects are designed and implemented and the objectives that are sought. For example, the capacity for a class of undergraduate students to collaborate on a course project with a community group with a clear set of needs that ensure that both the learning outcomes of the student and the needs of the community group are met is complex. It requires constant negotiation, renegotiation and managing sometimes contradictory objectives.

Neogeography in the Classroom

As neogeography and the geoweb have become increasingly established as an important area of research and development within the university, its role in the classroom has broadly taken two distinctive directions. The first is a computer science approach with a focus on developing requisite programming skills to work effectively in a geoweb environment. Increasingly, this approach requires a broad skill set including the specifics of working with online mapping application programming interfaces (APIs), as well as having an understanding of a range of web programming languages including PHP, Javascript, CSS, AJAX and a growing number of others. This computer science approach is rarely situated inside geography departments and the students are often not geography undergraduates. The second, very different approach, is to use an appropriate solution from the growing range of available, often open-source, applications and technologies that suit the specific objectives of a project. In this second approach, there is less of an emphasis on tool creation but more on application deployment. This is because the tool is intended to address spatially located issues where the map plays a role in creating a better understanding of these issues. In other words, we see the deployment of neogeography tools to address place-based issues, supported by the overall objectives of undergraduate teaching and learning. This second approach is more prevalent within geography departments.

This next section will describe the application of place-based education (PBE) as it relates to the teaching of a neogeography undergraduate course at UBC Okanagan campus. We use PBE as both a framework and a lens to reflect on the application of neogeography tools in the classroom as well as in the deployment of these tools by groups of undergraduate students to address locally bounded problems. Undergraduate students, working in both teams and individually, chose to use existing available technologies to support their projects.

Place-Based Education

Place-based education (PBE) has been broadly defined within multiple disciplines but is often used interchangeably with terms such as 'service learning', 'contextual learning', 'experiential learning', 'community-oriented education', 'bioregional education', 'civic education' and 'project-based education' (Gruenewald 2003; Powers 2004; Smith 2007). While place-based education was largely established within a rural elementary education framework (McInerney et al. 2011), the goals of a place-based pedagogy are to move beyond an individualistic perspective that encourages 'teaching to the test' and 'learning to earn', which have each become increasingly significant issues within current post-secondary settings (Gruenewald 2005; Semken and Freeman 2008). Inherently multidisciplinary, place-based learning allows students to become familiar with various aspects of local places which might have been previously overlooked in today's Internet-focused, globalized world. In place-based education, the focus of learning is on the local environment, so that learning experiences do not occur in isolation from the daily lives of students. Therefore, 'teaching is grounded in what students are familiar with; actualities rather than abstractions' (Lewthwaite 2007, 5).

To understand the multiple and overlapping factors that affect places (sociological, environmental, political, etc.), students are encouraged to become active but conscientious participants. Place-based education has a number of positive impacts on learners, including increased resourcefulness, independence and confidence as well as increased academic engagement and achievement (Schlottmann 2005). By engaging and interacting with local community members and organizations, students access diverse viewpoints and develop problem-solving and observational skills, meanwhile strengthening relationships with local organizations, politicians and community members (Powers 2004; Smith 2007).

Applying a critical lens to place-based education can be particularly valuable in post-secondary settings as a strategy for teaching students to 'read the world' so as to understand various local social and ecological injustices (Freire 1996). A critical pedagogical approach to PBE can be a way to empower students to become producers of knowledge (McInerney et al. 2011). The more students develop a 'sense of place' or place attachment, the more invested they become in their local community and thus move from being simply residents to inhabitants (Gruenewald 2003; Semken and Freeman 2008).[1]

Like any pedagogical approach, PBE is not without its challenges or critiques. Problematically, the term 'place' has been used in the PBE literature often without an accompanying definition and, as a result, has been critiqued for being under-theorized, due to 'prevailing assumptions about the notions of place, identity and difference' (McInerney et al. 2011, 9). Within the specific context of this chapter, we understand the term 'place' to mean the immediate and physical space of the

[1] According to Semken and Freeman (2008), 'a combined set of place meanings and place attachments, held by a person or a group, constitutes a functional definition of the sense of place' (1043).

City of Kelowna in which the university is located, as well as the region of the Okanagan Valley, and more broadly how we fit into the province of British Columbia. Yet we also draw on the work of Tuan (1990) and use the concept of place to understand that both undergraduate students and faculty members have well-formed attachments to this space based on lived experiences and feelings, in which they live, study and play. Furthermore we believe that by consciously engaging in a PBE approach with undergraduate geography students, they might further deepen their relationships to these places.

While PBE has been critiqued for being anti-universalist or detracting from students' abilities to think or work globally, the learning outcomes clearly demonstrate transferable real-world skills (McInerney et al. 2011). One of the main challenges with PBE is that as a method, it often lies outside of the standardized narrow schemes that are conventionally used to evaluate students (Gruenewald 2005; Smith 2007). As such it has been critiqued for being difficult to replicate and be statistically verifiable (in comparison to standardized testing) and can thus draw heightened surveillance from others – including administrators. Furthermore, due to the insular structure of post-secondary institutions, it can be difficult to foster authentic collaborative relationships with local communities that move beyond institutional public relations ploys. The most problematic challenge with a critical approach to PBE is that its problem-focused nature can cultivate a sense of hopelessness among students if not balanced by an investigation into what positive aspects of place need to be conserved and respected (Gruenewald 2003; Smith 2007).

There are three main components to PBE that position the theory as a useful approach within the context of teaching and learning neogeography: (1) the ability of a place-based approach to empower students to take responsibility for the roles that they can inhabit as place-makers and local citizens, (2) a profound investment in social and environmental care as a result of an increased 'sense of place' and (3) a strengthening of 'real-world' interpersonal, analytical and technical skills as well as academic achievement among students from diverse cultural and socio-economic backgrounds.

Empowering Place-Makers Through Civic Involvement

Promoting the interconnectedness between people and place, the discipline of geography has a strong history of citizenship education (Elfer 2011). Despite a tendency towards providing solutions for social problems, the link between the role of education as a mediator in the production of space to students as place-makers (or citizens) is rarely made. According to Gruenewald (2003), this relationship is a central focus of a critical pedagogy of place, which more specifically aims to '(a) identify, recover, and create material spaces and places that teach us how to live well in our total environments (reinhabitation); and (b) identify and change ways of thinking that injure and exploit other people and places (decolonization)' (9). With the ultimate goals of reinhabitation and decolonization, Gruenewald advocates for an approach that critically examines local ecological relationships and cultural politics so as to determine what aspects of place are problematic, thus requiring transformation and, equally important, what aspects of place are valuable and should be conserved.

Through working with local community members and organizations, students access viewpoints that are not only focused on environmental care but also include political, sociological and economic perspectives. For Smith (2007), this had a profound effect on how students perceived their own role in terms of civic involvement: 'They came to perceive themselves as citizens capable of participating in public conversations that could protect and improve their own lives and the lives of those around them. They and others began to recognize their voices as significant and as potentially powerful as the voices of any other participants in civic government' (195). Similarly, for Powers (2004), the consistent implementation of community-oriented service-learning experiences coupled with continued dialogue between students and teachers resulted in not only increased participation in local matters but also enhanced academic performance. Powers argues that such civic involvement results in the broadening and deepening of social capital – 'the invisible web of relationships…such as networks, norms, and social trust that facilitate coordination and cooperation for mutual benefit' (19).

Developing relationships with local community members and organizations as well as relationships to the local environment further nurtures student commitment to local issues but can also develop an awareness of places and knowledges beyond the scope of the local and familiar. As a decolonizing process, which is defined by Gruenewald (2003) as 'a metaphor for the process of recognizing and dislodging dominant ideas, assumptions and ideologies as externally imposed' (71), a critical place-based approach encourages students to move beyond celebrating their own lived experiences to a critical dialogue on relationships between place, power and privilege (McInerney et al. 2011). Such an approach has a tendency to orient students towards issues of social justice, especially in terms of ecology (eco-justice).

Encouraging Eco-justice and Reinhabitation

A tangible strength of a place-based education approach is the role that it plays in teaching students about achieving eco-justice. As students develop an increased sense of place or place attachment, they become invested in the ecological impacts that negatively affect significant places (Semken and Freeman 2008). According to Gruenewald, an eco-justice approach is based on the following aims:

1. Understanding the relationships between ecological and cultural systems, specifically, between the domination of nature and the domination of oppressed groups
2. Addressing environmental racism, including the geographical dimension of social injustice and environmental pollution

3. Revitalizing the non-commodified traditions of different racial and ethnic groups and communities, especially those traditions that support ecological sustainability
4. Reconceiving and adapting our lifestyles in ways that will not jeopardize the environment for future generations (2003, 6)

Central to the goals of eco-justice within PBE is reinhabitation or living well within a place that may be ecologically or socially injured. Such inhabitance is described as ‘art requiring detailed knowledge of a place, the capacity for observation, and a sense of care and rootedness’ (9). Promoting local ecological sustainability over resource extraction and competition, a main focus within the eco-justice discourse of PBE has been in preserving ‘the commons’ (Bowers 2008). While this involves learning about local environmental problems, the focus on conservation includes ‘an awareness of the ecological importance of the many forms of intergenerational knowledge, skills and patterns of interdependence and support that can also be understood as traditions’ (Bowers 2008, 329). Taking a place-based approach to eco-justice is particularly valuable for those who might not easily identify with environmentalism, as they can still value and advocate for the places where they live (Gruenewald 2005).

Developing ‘Real-World’ Skills

Despite the critique of being too provincial, a locally focused place-based approach has the potential to teach students real-world skills that can be transferred to other sites and circumstances. Breaking down the barrier between education and ‘real life’, students work collaboratively with local community members to solve specific problems. Such problem-solving not only requires a reflexive process, whereby students learn to recognize the ways in which their own perspectives have been shaped by various social processes, but also involves the application of a variety of skills including resourcefulness, active listening, problem-solving, observation, public speaking and, of course, interpersonal skills. As a place-based approach typically involves a collaborative process, group members often learn technical skills from one another, which are then developed through the experiential process.

Neogeography Learning in and About Place

Through a series of readings, lectures, seminars and a term project, we taught a course that critically explored the evolution of neogeography and the geoweb, its increasing prevalence and importance in society, as well as its ability to influence the manner in which students consider our relationship to space, place and the world around us, with a particular focus on the ways in which we can contribute to solving local challenges. The course ran in the winter semester of 2014, and 15 students were enrolled. The course drew on seminal texts from the discipline of geography, GIScience (including critical GIS), sociology and Applied Internet Research. A central focus of the course was a term project that, in most cases, involved the direct involvement/engagement of undergraduate students with members of the public using neogeography tools. Breaking the mould of traditional undergraduate GIT courses where students tend to develop skills related to specific software packages, we encouraged students to instead explore the range of neogeography tools and applications to match their project objectives. Furthermore, considering the scope of the course, student projects were required to be interdisciplinary, include a spatial component as well as reflect an appreciation for the theoretical concepts behind both geography broadly and neogeography specifically.

Moving beyond understandings of Cartesian elements of geoweb applications, it was pertinent that students explored core theoretical principles including constructs such as place, space, community and the environment. Undoubtedly theory transformed into practice, as the ways in which we discussed and critically reflected on the projects resulted in a general gravitation towards problem-solving-related projects focused on issues of social justice. An evaluation rubric was created collectively with the students in order to clarify the evaluation criteria for the project component of the course.

It should be noted that although the projects were initiated and developed by the students, from an instructor's experience, this approach required a greater time and energy commitment, not less, to support the students. Firstly, instructors had to put resources into place to support the projects (e.g. making available the time of a paid geoweb programmer, providing access to software licences and equipment). Secondly, instructors had to be active members of the class, constantly monitoring the development of projects, communicating with and supporting the aspirations of students. Furthermore, the instructor involvement in the projects seeped out of classroom and drew on our own social network and capital by making connections with colleagues and members of the public or linking students to communication relations in the university, not as a gatekeeper but as an active collaborator in these projects. But as a result of this close proximity to the students and their projects, we were also invested in their outcome and success.

The next sections will describe two projects, of the seven undertaken by the class, that provide clear examples of the significance of PBE in both developing and implementing these projects as well as understanding some of their outcomes. The first is the Place and Pipelines project, an online map designed to crowdsource information from the public related to the proposed Enbridge pipeline through Northern British Columbia, the second is the development of an Okanagan Fruit Tree Project (OFTP) map that was designed to support a local NGO food justice organization (by the same name) in planning its logistical activities.

Place and Pipelines

The Northern Gateway Pipeline is a proposal to build a dual pipeline that connects Alberta's Industrial Heartland with the coast of British Columbia. Tankers will bring petroleum condensate to the port of Kitimat, BC, where it will be pumped to the pipeline terminus near Bruderheim, Alberta, at a rate of 192 thousand barrels per day. Under the Canadian Environmental Assessment Act, any large industrial project like Northern Gateway requires an Environmental Assessment (EA) before it can proceed. In the case of the Northern Gateway Pipeline, federal regulators chose to use the Joint Review Panel format to conduct the EA. This process provides a forum for industry experts, NGOs and local residents to provide feedback about the proposal, and the proponent (Enbridge Northern Gateway Pipelines Inc.) has the opportunity to rebut testimony. All testimony and correspondence are transcribed and become a part of the public record. At the end of the process, the panel makes a recommendation to the federal government based on the information presented. A recommendation to approve the project with 299 conditions was made in December 2013. Throughout the hearing process, and since the subsequent project approval, there has been a significant public outcry opposing the construction of the pipeline and the associated increase in tanker traffic off the coast of BC.

Andrew Barton, a geography undergraduate student, as part of another research project he carried out in 2012, collected the material used in the Place and Pipelines project. That research included a thematic analysis of the transcripts from the Joint Review Panel hearings and fieldwork that resulted in a photographic essay and narrative of his journey through the region. Barton created an interactive geoweb mapping project that spatially presented this data, as well as invited members of the public to voluntarily crowdsource their own spatially encoded photos, videos and text to describe and represent their own experiences with the region that the pipeline is intended to cross. Once their data was included on the map, users could visually compare the direction and location of pipeline with their own experiences, thus bringing and influencing their own thoughts and ideas about the human and environmental impacts of pipeline. The mapping platform used for this project is Geolive, an online participatory mapping tool that has been developed at UBC Okanagan since 2009.

Initially, the Geolive map was populated using data collected from Barton's previous research. He created custom markers for the different data types. The markers represent quotes from the Joint Review Panel, the hearing locations and photos from his journey along the proposed route. Clicking on a Quote marker shows a quote from the hearings, with a cross-reference to the original government document. Clicking on the hearing marker opens up a box that contains cross-referenced information on the hearings that were held in that location and the associated transcripts. The site provided a spatially referenced visual and written tool that presents resources concerning the Northern Gateway Environmental Assessment hearing process in a visually engaging manner (see Fig. 1).

Fig. 1 Place and Pipelines project

A second component of the research project transformed the Place and Pipelines project into a participatory mapping project, by adding the ability for the website to display geographic information volunteered by members of the public. Users can login to the geoweb map using their existing Facebook or Google Plus account, as well as by creating a new account on the system. Once this is done, new marker options appear on the right-hand side of the browser window, which can then be dragged onto the map and annotated with text, photos, video and audio. The markers can be positioned visually on the map or by entering an address into the location field.

Okanagan Fruit Tree Project

In recent years, Kelowna has become a regional centre for alternative food initiatives and a 'full-fledged foodie destination' (Michaels 2014). This transformation is evidenced by the increasing number of grocery stores and small business models selling often high-priced, local and organic food, the steady rise of wine culture and the farm-to-table concept in which diners can indulge in high-priced dishes made from ingredients traced back to local farms and orchards (Michaels 2014). While these initiatives indicate the growing presence of an alternative food system that offers new ways to access food in Kelowna, and more broadly the Okanagan Valley, these initiatives primarily target a niche demographic, catering to the palates of middle- and upper-class citizens and tourists. Meanwhile access to sufficient, safe and nutritious food continues to be a critical issue for many: food bank usage has grown at an annual rate of 4% in Kelowna and West Kelowna, while the national rate has grown by 1% (Jeffery 2014). Over 4000 people from Kelowna and West Kelowna relied on the food bank to meet their dietary needs in March 2014, and over one-third of those users were children (Jeffery 2014). Such alarming statistics point out the socially unjust and exclusionary spaces within the alternative food paradigm.

Despite the exclusion and unequal access created by many of the niche, alternative food initiatives, there are instances where a food justice agenda is being used to engage citizens and facilitate access to food. One such example comes from the work of the Okanagan Fruit Tree Project (OFTP), a local non-profit organization that gleans, or harvests, backyard fruit trees to reallocate nutritious produce to those in need. The OFTP is a highly politicized endeavour that actively seeks to work with people excluded from the Okanagan's niche food system and enable them to access inexpensive and healthy foods while encouraging them to critically engage in issues of food security within the region. In the second project, a team of six students worked with members of the OFTP to develop an interactive geoweb-based map to improve the organization's scheduling and picking efficiency. The specific objectives of the project, determined collaboratively by both the students and members of OFTP, were to develop an interactive mapping website that could act as a marketing tool and allow more people in the community to learn about and engage with the project. They also wanted to provide OFTP with a tool that would help streamline their day-to-day operations and data management. This entailed transitioning OFTP's data from a series of Microsoft Excel spreadsheets to a geoweb tool that would spatially encode the data and then present it in a way that made it more useful to both the organization's administrators and also the volunteer fruit pickers.

Similar to Andrew Barton's Place and Pipelines project described above, the team chose to use Geolive as the geoweb application to capture, manage and communicate the OFTP's data. Specifically, the OFTP team were interested in leveraging Geolive's ability to manage the online map's users as groups with different access permissions. The OFTP required two interactive layers – one openly accessible to the general public and the other only visible to the project's administrators. The public layer provides information regarding maintenance of garden and fruit trees, how to receive produce through the OFTP, where donations can be made, relevant and upcoming public events and how to get involved. This information is accessible to members of the public via makers which are placed on the Geolive map. The private layer contains information with the street addresses and contact details of fruit tree owners, the fruit type, the dates when trees were picked, who picked them, the yield and how the tree or garden is managed (spray, no spray, etc.) (see Fig. 2). As an active management tool, this layer provides OFTP administrators with not only the opportunity to add new information and plan their annual picking

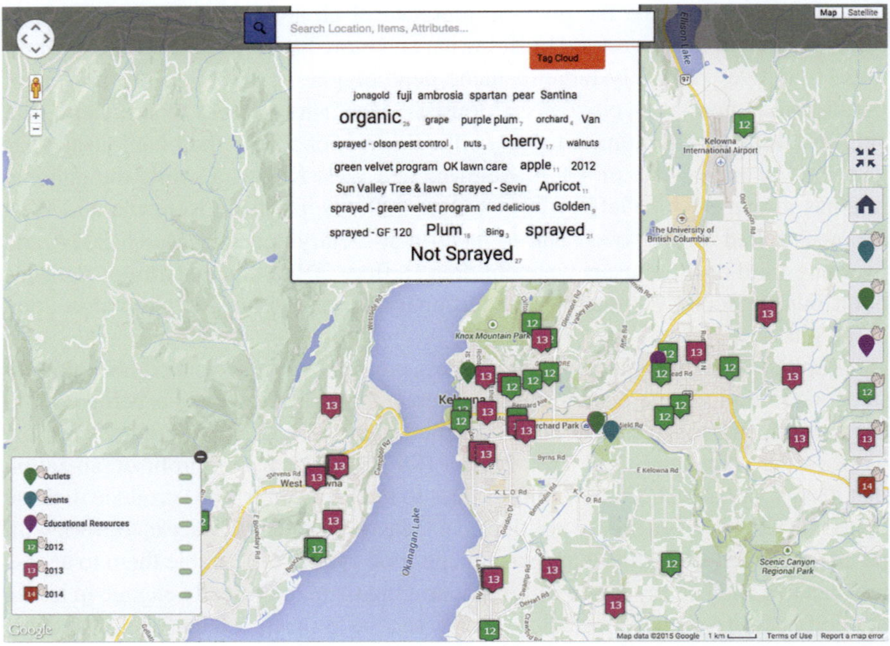

Fig. 2 Okanagan Fruit Tree Project map

regime, but it also provides a visualization of the project's future expansion into other areas of the Okanagan.

Discussion

To frame our reflections of the two participatory geoweb projects described in this chapter, we will use the three main components of PBE identified. In particular, we are interested in how a PBE approach to teaching and learning of neogeography can firstly contribute to creating more civic-minded and engaged undergraduate students; secondly, whom are prepared to address issues of eco and more broadly social justice; and thirdly, to support student acquisition of real-world skills.

Empowering Place-Making and Civic Involvement

The Place and Pipelines project focused on drawing the public into a geoweb-based discussion related to a large-scale development that could potentially have significant economic and ecological impacts. The project was publicized in a number of ways, including radio, news releases, mailing lists, social media and social networks. As a result, interest in the project extended beyond local populations. Between January 15 and May 23, 2014, Google Analytics tracking was enabled to determine website interactions and referral sources. Out of the 4282 page views (1238 unique sessions), only 239 of these sessions originated from Kelowna (the project's base), and these might even be disregarded as they likely originated from people directly involved with the project. The remaining 999 sessions mainly originated in Canada, with the top ten cities being Vancouver, Victoria, Edmonton, Toronto, Calgary, Richmond, Ottawa, Prince George, Vernon and Surrey. Visits from North America totalled 953 (Kelowna numbers removed), with South America being the next largest originator at only 12 visits. These statistics demonstrate the extent of the project's reach and the broad impact that an undergraduate project can have in terms of drawing civic attention to a pressing provincial issue.

Directly relating to the goals of place-based education, the Place and Pipelines project allowed members of the public to access and share a diversity of viewpoints that extended beyond environmental concern but also included economic, sociological and political perspectives. Overall this was a mature and non-partisan mechanism of enabling members of the public to engage on this contentious issue. Although Barton seeded the map with photographs and direct quotes from the Joint Review Panel, he ultimately sought to represent the experiences and views of members of the public. Barton was less interested in sharing his own perspectives and knowledge but rather focused on providing a platform for others to share their own and thus to facilitate members of the public to engage and reflect on the human impacts of the issue. In other words, similar to the principles of participatory mapping, he does not position himself as 'expert' but as a facilitator, by providing the means for others to contribute as experts of their own experiences.

Similarly, the OFTP interactive map resulted in the proactive and direct engagement of the undergraduate team with an issue of local social and economic significance. As a tangible marketing platform, the interactive map aimed to assist in addressing food security issues in the Okanagan and help contribute to overcoming this issue using the Geolive participatory mapping platform. Through developing a map that manages, organized and visualizes data, a space was created for the public to engage with the issue. Furthermore the tool enhanced the goals and values of the OFTP, the interactive map helped streamline operations by developing a way for organizers to manage logistics while also planning for the future. Through promoting community engagement, public awareness and volunteerism, the OFTP map empowered students and thus feel that they were having a direct impact on an issue that affects their local community. Furthermore, through collaborating with the OFTP organizers to develop a useful platform, students not only broadened their social network (and thus, social capital) but, similar to the Places and Pipelines project, were provided the opportunity to become recognized supporters of social justice issues in both a civic context and by the university.

Both projects described in this chapter were different in terms of their specific relationships with community members and organizations (and by extension, place). For instance, the students built on a long-standing pre-existing relationship between an instructor and the OFTP but became increasingly invested in the effectiveness and successful use of the tool that they created. In contrast, Barton's project did not involve a relationship with a specific identified organization but instead invited the general public to contribute their experiences with the hope of tapping into a pressing debate in democratic way. Through the site, Barton encouraged people to engage with the issues; however, he did not privilege a set of specific viewpoints – rather he hoped to encourage dialogue.

However, in both examples we believe that through adopting a PBE approach to teaching neogeography, undergraduate students felt as if they are contributing something tangible and positive to a place to which they are living, perhaps born in, or otherwise invested in. This opportunity to act as an effective citizen in that place greatly increased their interest and motivation in the study. We posit that this is because they feel more invested, both physically and emotionally, to the issues and to the potential benefits that can be derived from their deployment of geoweb technologies to directly address local challenges. This is very different from undergraduate projects based on hypothetical scenarios, using predefined datasets, with limited outcomes.

Furthermore, we believe that a PBE approach actively transforms student relationships to the places and issues addressed by their projects. For example, the team working on the OFTP map knew little about food justice and food access in the Okanagan, but through engaging directly with these issues in their own community, they became more aware of the inequalities that exist around them. This has the potential to have a profound impact of student development, as students view their role in the project as a sort of community volunteerism.

Eco-justice

The second aim of place-based education involves a focus on eco-justice and developing sense of place or place attachment. This is central to the Place and Pipelines project as it was initiated as a response to the potential negative environmental effects that could result from the pipelines and the lack of public representation in the issue. Through crowdsourcing, the project sought to display the relationship between those people affected by the ecological impacts of the pipeline projects. As the public voiced their opinions and experiences, it was evident that their concerns were centred on geographic areas that were of ecological importance.

Significantly, the OFTP interactive map also sought to connect people to the environment through geographical representation. Through turning private resources (fruit) into public ones for the common good of enhanced food security and reduced waste, learning about the OFTP was a lesson in eco-justice and sustainability. Motivated by the aims of the OFTP , student engagement with the project reinforced their sense of place or attachment to the Okanagan and those who inhabit the area. Understanding that the Okanagan is 'socially injured' in that there are many people who cannot readily access healthy food, students worked to engage volunteers who could assist in caring and advocating for their local community.

However, we need to be realistic about the effectiveness of these projects in supporting meaningful change. Despite students hoping that their projects will have a profound environmental or social impact, the potential for success is often hampered by the limitations of being an undergraduate student. Project life spans are built around limited time periods imposed by the semester system. This greatly constrains the available time to achieve traction and support for the work and even more importantly brings into question the longevity and thus sustainability of these projects. As students are often temporary members of the university community, there is a need to recognize the ways in which student limitations impact the ability of projects to support long-term effectiveness and thus social change. Furthermore, community partners who understand these limitations of students' time are more cautious in investing their support into these projects because they are aware that long-term assistance is often limited. These are ethical issues that need careful consideration by students and instructors before commencing any such PBE project.

Real-World Skills

A focus of the PBE framework is the acquisition and development of 'real-world' skills. Unlike many traditional undergraduate GIS courses, the neogeography course focused on the development of both technical and social skills. We recognized at the outset of the course that many undergraduate geography students have limited or, in some cases, no computer programming background, which makes mastering the technical aspects of the geoweb very difficult. To overcome these limitations, we used Geolive to provide a visual, menu driven and straightforward way for students to develop and manage their online maps without having any previous knowledge of computer programming. Students also relied on the Geolive head programmer to set up their project website and write any specific code that would be required to support the individual projects. Nonetheless, students were required to develop image editing skills to create the website design, logos, map-markers and icons. Furthermore, students were also required to create and manage map layers and tag clouds, take responsibility for website monitoring and moderation and manage the contributed data. Both projects ran into limitations around student programming capacity and required a level of technical support. However, it is important to note that members of both projects were realistic in terms of understanding their technical limitations and demonstrated great resourcefulness in searching out solutions to the various challenges that arose.

Although the projects involved developing specific technical skills, we recognize that these hard skills were less significant than the soft skills acquired by students throughout the duration of the course. In terms of interpersonal and social skills development, the students were required to develop rapport with community members, which required conversational, observational, problem-solving and active listening skills. Students also had the opportunity to engage in group-based decision-making and networking. A significant factor of each project was the public relations component, which required social media literacy, creative design skills and organizational skills as well as the ability to develop relevant and appealing content. In the case of the Place and Pipelines project, Barton was involved in a number of both video and radio interviews. For members of the OFTP team, they presented their website and outcomes back to members of the OFTP board of directors, at which time they also presented training manuals to ensure the longer-term use of the site.

Other soft skills included team organization, delegation around different tasks and building on individual strengths. For example, in the OFTP team, two of the student team assumed responsibility for project communications and information gathering, two were responsible for data management and inputting information onto the map, and the two most technically competent designed the site using the administrative backend of Geolive. While students played to their individual strengths, the project succeeded as a result of their collective motivation to support the OFTP. Not only were student ambitions much higher than those of the instructors, but as a result of the meaningful relationship that they had developed with the OFTP, students went above and beyond expectations as they were truly invested in providing a useful tool for improving food security in the Okanagan.

Conclusions

Returning to the original pedagogic objectives of the neogeography course, we encouraged students to take ownership over a project that focused on using participatory geospatial tools to address place-based challenges through making the relationship between land and communities more obvious. Although the projects began as a simple course assignment, through the students own agency and dedication, they took an active, hands-on approach, which in turn rendered each project to be of greater significance for the students. We firmly believe that by addressing locally important issues and critically engaging with place, there was a significant increase in the level of motivation and engagement of students.

Although significant skills were developed through engaging in place-based learning using geoweb tools, this approach is embedded in a much larger politic and reflection by students. These critical lenses directly complement the deployment of participatory geospatial tools but also encourage students to consider the complexity of working with communities, the embedded politics of the work and research that they do and the potential impacts (both positive and negative) of their work. It is important to note that, in a number of cases, students noted that this PBE approach was transformative moment in their undergraduate experience. This is clearly reflected by the large number of students in the course that continue to ask about the long-term uptake and use of their projects; this was particularly the case with the OFTP team.

Although both projects described in this chapter represented a resistance to the structural problems that exist within each context, environmental damage in the province and food security challenges in the Okanagan, they also differed in many ways. We feel that a key difference between the objectives of the OFTP map and the Place and Pipelines project is the issue of scale and the intended audience for the map. The OFTP set out to specifically support one local organization through directly providing a mapping resource to address local issues of food security. As such, the scale is very localized and the audience defined also by that geography. Barton's Place and Pipeline map had a more ambitious scale (the provincial level) and a less clearly targeted audience. In many ways through creating a public platform for dialogue, it sought a broader potential impact and reach, but also because it lacked a partnering organization, it is more prone to lacking longevity and impact.

On a final note, we would also like to identify that as universities move towards a greater sense of commitment to engaging in research that addresses community needs and challenges, these types of PBE undergraduate projects that make use of participatory mapping, the geoweb and other geosocial tools represent a quick win for universities, as well as a meaningful learning experience for the students. However, we should not downplay the importance of these types of project to contribute to building community cohesion through raising community awareness, as well as the broader university student base, to engage in place-related issues and to raise awareness about pressing land-related issues, and ultimately contribute to empowering individuals to take a greater role in seeking solutions to these issues.

Acknowledgments The authors would like to thank Andrew Barton, author of the Places and Pipelines project, as well as Travis Eek, Stevie Frew, Liz Waterfield and Alex Rutledge, the creators of the Okanagan Fruit Tree Project web map. We would also like to acknowledge Casey Hamilton and Ailsa Beischer from the Okanagan Fruit Tree Project for their input, ideas and support of the geoweb project. Finally we would like to thank Nick Blackwell, the head Geolive programmer, and Joanne Carey, the Centre for Social, Spatial and Economic Justice administrator.

References

Bowers, A. (2008). Why a critical pedagogy of place is an oxymoron. *Environmental Education Research, 14*(3), 325–335.

Brown, G., & Raymond, C. M. (2014). Methods for identifying land use conflict potential using participatory mapping. *Landscape and Urban Planning, 122*, 196–208.

Corbett, J. (2009). *Good practices in participatory mapping*. Rome: International Fund for Agricultural Development.

Coulson, M. R. C. (1977). Political truth and the graphic image. *The Canadian Cartographer, 14*(2), 101–111.

Craig, W. J., & Elwood, S. (1998). How and why community groups use maps and geographic information. *Cartography and Geographic Information Systems, 25*(2), 95–104.

Crutcher, M., & Zook, M. (2009). Placemarks and waterlines: Racialized cyberscapes in post-Katrina Google Earth. *Geoforum, 40*(4), 523–534.

Dorling, D. (1998). Human cartography: When it is good to map. *Environment and Planning A, 30*, 277–288.

Elfer, C. A. (2011). *Place-based education: A review of historical precedents in theory and practice*. PhD dissertation, University of Georgia, Athens.

Elwood, S. (2002). GIS use in community planning: A multidimensional analysis of empowerment. *Environment and Planning A, 34*(5), 905–922.
Elwood, S. (2008). Volunteered geographic information: Key questions, concepts and methods to guide emerging research and practice. *GeoJournal, 72*(3–4), 133–135.
Elwood, S., & Ghose, R. (2001). PPGIS in community development planning: Framing the organizational context. *Cartographica: The International Journal for Geographic Information and Geovisualization, 38*(3), 19–33.
Elwood, S., & Leszczynski, A. (2011). Privacy, reconsidered: New representations, data practices, and the geoweb. *Geoforum, 42*(1), 6–15.
Elwood, S., Schuurman, N., & Wilson, M. (2011). Critical GIS. In *The Sage handbook of GIS and society* (pp. 87–106). London: Sage.
Freire, P. (1996). *Pedagogy of the oppressed*. London: Penguin Books.
Ghose, R. (2007). Politics of scale and networks of association in public participation GIS. *Environment and Planning A, 39*(8), 1961–1980.
Goodchild, M. F. (2007). Citizens as sensors: The world of volunteered geography. *GeoJournal, 69*(4), 211–221.
Gruenewald, D. A. (2003). Foundations of place: A multidisciplinary framework for place-conscious education. *American Educational Research Journal, 40*(3), 619–654.
Gruenewald, D. A. (2005). Accountability and collaboration: Institutional barriers and strategic pathways for place-based education. *Ethics, Place & Environment, 8*(3), 261–283.
Haklay, M., Singleton, A., & Parker, C. (2008). Web mapping 2.0: The neogeography of the Geoweb. *Geography Compass, 2*(6), 2011–2039.
Hall, G. B., Chipeniuk, R., Feick, R. D., Leahy, M. G., & Deparday, V. (2010). Community-based production of geographic information using open source software and Web 2.0. *International Journal of Geographical Information Science, 24*(5), 761–781.
Harley, J. B. (1989). Deconstructing the map. *Cartographica: The International Journal for Geographic Information and Geovisualization, 26*(2), 1–20.
Jankowski, P. (2011). Designing public participation geographic information systems. In T. Nyerges, H. Couclelis, & R. McMaster (Eds.), *The Sage handbook of GIS and society* (p. 347). London/Thousand Oaks: Sage.
Jeffery, C. (2014, November 5). More hungry mouths to feed this year in the Okanagan. *Kelowna Now.* Available from: http://www.kelownanow.com/columns/promotion/news/Promotion/14/11/05/More_Hungry_Mouths_to_Feed_This_Year_in_the_Okanagan. 5 November 2014.
Johnson, P. A., & Sieber, R. E. (2011). Motivations driving government adoption of the geoweb. *GeoJournal*, 1–14.
Keen, A. (2007). *The cult of the amateur: How blogs, MySpace, YouTube, and the rest of today's user-generated media are destroying our economy, our culture, and our values*. New York: Broadway Business.
Kienberger, S. (2014). Participatory mapping of flood hazard risk in Munamicua, District of Búzi, Mozambique. *Journal of Maps*, 1–7.
Kingston, R. (2011). Online public participation GIS for spatial planning. In T. Nyerges, H. Couclelis, & R. McMaster (Eds.), *The Sage handbook of GIS and society* (p. 361). London/Thousand Oaks: Sage.
Kwan, M. P. (2002). Feminist visualization: re-envisioning GIS as a method in feminist geographic research. Annals of the Association of American Geographers, 92(4), 645–661.
Laituri, M. (2003). The issue of access: An assessment guide for evaluating public participation geographic information science case studies. *Urisa Journal, 15*(2), 25–32.
Leitner, H., McMaster, R. B., Elwood, S., McMaster, S., & Sheppard, E. (2002). Models for making GIS available to community organizations: Dimensions of difference and appropriateness. In W. Craig, T. Harris, & D. Weiner (Eds.), *Community participation and geographic information systems* (pp. 37–52). London: Taylor and Francis.

Lewthwaite, B. (2007). From school to community-based school: The influence of an aboriginal principal on culture-based school development. *Canadian Journal of Educational Administration and Policy*, (64), 1–18.
Livingstone, D. (1992). *The geographers tradition*. Oxford: Blackwells.
Macintyre, B. (2008, March 14). Whatever next: A map of Utopia – Thanks to the internet we are living in a golden age of cartography. *Times of London*.
McCall, K., Martínez, J. A., & Verplanke J. J. (2014). Shifting boundaries of volunteered geographic information systems and modalities: learning from PGIS. *ACME: An International E–journal for Critical Geographies: Open Access*, 36.
McInerney, P., Smyth, J., & Down, B. (2011). 'Coming to a place near you?' The politics and possibilities of a critical pedagogy of place-based education. *Asia-Pacific Journal of Teacher Education, 39*(1), 3–16.
Michaels, K. (2014). Close up: Kelowna farm to table movement flourishing. *Kelowna Capital News*. Available from: http://www.kelownacapnews.com/news/265833601.html. 4 November 2014.
Panek, J. (2014). How participatory mapping can drive community empowerment--a case study of Koffiekraal, South Africa. *South African Geographical Journal* (ahead-of-print), 1–13.
Parker, B. (2006). Constructing community through maps? Power and praxis in community mapping. *The Professional Geographer, 58*(4), 470–484.
Poole, P. (1995). Geomatics, who needs it? *Cultural Survival Quarterly, 18*, 1–77.
Powers, A. L. (2004). An evaluation of four place-based education programs. *Journal of Environmental Education, 35*(4), 17–32.
Rambaldi, G., Chambers, R., McCall, M., & Fox, J. (2006). Practical ethics for PGIS practitioners, facilitators, technology intermediaries and researchers. *Participatory Learning and Action, 54*(1), 106–113.
Scharl, A., & Tochtermann, K. (2007). *The geospatial web: How geobrowsers, social software and the web 2.0 are shaping the network society*. London: Springer.
Schlottmann, C. (2005). Introduction: Place-based and environmental education. *Ethics, Place & Environment, 8*(3), 257–259.
Semken, S., & Freeman, C. B. (2008). Sense of place in the practice and assessment of place-based science teaching. *Science Education, 92*(6), 1042–1057.
Sieber, R. E. (2006). Public participation geographic information systems: A literature review and framework. *Annals of the Association of American Geographers, 96*(3), 491–507.
Sieber, R. E. (2007). Spatial data access by the grassroots. *Cartography and Geographic Information Science, 34*(1), 47–62.
Smith, G. A. (2007). Place-based education: Breaking through the constraining regularities of public school. *Environmental Education Research, 13*(2), 189–207.
Stephens, M. (2013). Gender and the GeoWeb: Divisions in the production of user-generated cartographic information. *GeoJournal, 78*(6), 981–996.
Sui, D., & DeLyser, D. (2012). Crossing the qualitative-quantitative chasm I Hybrid geographies, the spatial turn, and volunteered geographic information (VGI). *Progress in Human Geography, 36*(1), 111–124.
Sui, D. Z., & Holt, J. B. (2008). Visualizing and analysing public-health data using value-by-area cartograms: Toward a new synthetic framework. *Cartographica, 43*(1), 3–20.
Tuan, Y.-F. (1990). Topophilia: A Study of Environmental Perceptions, Attitudes, and Values. Columbia University Press.
Tulloch, D. (2008). Is VGI participation? From vernal pools to video games. *GeoJournal, 72*, 161–171.
Turner, A. (2006). *Introduction to neogeography*. Sebastopol: O'Reilly Media, Inc.
Warf, B., & Sui, D. (2010). From GIS to neogeography: Ontological implications and theories of truth. *Annals of GIS, 16*(4), 197–209.
Wood, D. (1992). *The power of maps*. New York: The Guildford Press.

Navigating Employment Prospects for New Graduates in the Geospatial Sciences

Niem Tu Huynh and G. Brent Hall

Abstract The ubiquity of geospatial science instruction has meant that basic geospatial knowledge is widespread. However, this is now no longer enough, relative to a decade or more ago, for students to attain related employment upon graduation. Even the best students are currently struggling to find appropriate work, and many are either deferring entering the job market by extending their education through to graduate degrees or attending technical finishing programs at colleges in order to gain deeper experience with advanced technical software use. Independent of the specific student strategy adopted, soft skills, networking, and school-based work experiences are complementary and equally important aspects of the process of transitioning from postsecondary education into a career. This chapter places into context the student decision of whether or not to extend study, as well as the institutional quandary of how best to position students for the transition to work. As such, the authors focus on hard and soft skills that are necessary to thrive within the current geospatial technology field. The discussion borrows from geospatial technology competency models as well as workplace surveys. The chapter concludes with a discussion of how the elements of the post-education employment puzzle can be pieced together.

Keywords Geospatial sciences · Geography careers · Soft and hard skills · Career path · Professional geographer

---

The views expressed in this chapter are expressly those of the authors and do not represent the views of their employers.

N. T. Huynh (✉)
Independent scholar, Montréal, QC, Canada
e-mail: n.huynh@esperaneuro.com

G. B. Hall
ESRI, Toronto, ON, Canada
e-mail: bhall@esri.ca

S. Balram, J. Boxall (eds.), *GIScience Teaching and Learning Perspectives*, Advances in Geographic Information Science,
https://doi.org/10.1007/978-3-030-06058-9_9

Introduction

The discipline of geography bridges the social and physical sciences, and this broad interdisciplinarity can potentially help to prepare students for employment in a range of occupations (Schlemper et al. 2014; Solem et al. 2013). Adding to its broad-based appeal, foundational concepts of the discipline such as location and spatial differentiation have found particular expression through the revolution of geospatial technologies and their widespread use.[1] In fact, current attention to the usefulness of maps has been "discovered" by members of the public (e.g., Esri's ArcGIS Online, Google Maps, Wikimapia, etc.), businesses (e.g., evaluating customer trade areas for store expansion planning and advertising), humanitarian efforts (e.g., Humanitarian OpenStreetMap), government (e.g., election results by district), and city planners and managers, among many other map users. This implicit awareness of the importance of geography and geospatial science is growing continuously.

As a result of the associated global growth of the geospatial sector, most countries, including Canada, report increases in the general demand for knowledgeable geospatial information workers. The Canadian National Occupational Classification in 2016 summarized a list of 80-plus job titles[2] for the general area of mapping, noting that these positions "are employed by all levels of government, the armed forces, utilities, mapping, computer software, forestry, architectural, engineering and consulting firms and other related establishments." Typically, mapping positions fall into one of five general levels, ranging from entry-level technician and analyst to programmer, coordinator, and, at the most senior level, positions such as project manager (GeoCommunity 2014).

Statistics suggest that 58,400 people in Canada are employed in a mapping-related position[3] (Employment and Social Development Canada 2016). Using available data from Quebec as an example, Job Futures Quebec[4] (2015) estimated that the majority of mapping positions are in public administration (50%) followed by professional, scientific, and technical services (33%) and private industry (e.g., architects, engineers and related services firms) (27%). The profile of employees in Quebec is dominated by males (68.4%), aged primarily between 25 and 44 years (57.7%). This occupation class has a high rate of full-time workers (97.2%) and an average income of $49,300 per annum.

The medium-term projection in Canada (2013–2022) for mapping-related jobs is reported to be one of excess demand, where openings are expected to be 16,340

---

[1] Penn State Public Broadcasting produced an engaging series of videos about Geospatial Revolution: http://geospatialrevolution.psu.edu

[2] Example titles of Mapping and Related Technologists and Technicians: http://cnp.edsc.gc.ca/English/NOC/ProfileQuickSearch.aspx?ver=16&val65=2255&val=2&val1=2255

[3] For statistical purposes, mapping and related technologists and technicians are classified with four other occupations under the umbrella title "Technical Occupations In Architecture, Drafting, Surveying And Mapping" by the Canadian Occupation Projection System.

[4] Jobs Future Quebec: http://www.servicecanada.gc.ca/eng/qc/job_futures/job_futures.shtml

compared to 10,998 job seekers by the end of the projection period (Employment and Social Development Canada 2016). Further, national surveys of Canadian employers (e.g., CCCE 2014a) and independent research of job advertisements (e.g., Borwein 2014a) suggest a shortage of workers in the general field of information technology (e.g., including Web developers and Web specialists), which is one area where geospatial science students can find work. Hence, the prospects for geospatial job entrants in the Canadian economy seem to be, at least based on these statistics, relatively promising for graduates. However, reconciling economic projections with the current *reality* for many geospatial sector job seekers leaves a number of unanswered questions (Helyer and Lee 2014).

In relation to the projected growth in labor demand, students (and their parents) increasingly see education, and especially higher education, as a way to gain relevant knowledge and skills that will lead to a future career. This view of the value of higher education represents a shift in mindset, relative to the past, away from traditional learning goals toward the more singular goal of job preparation. A recent survey of Canadian students found that 91% agreed that their motivation to attend university was indeed to "get a good job," ahead of the desire to learn (Canadian University Survey Consortium 2016). In the geospatial sciences, this shift has had an important influence on both the nature and availability of related education programs. In particular, many university and community college programs have responded to the reported market growth statistics by branching out from traditional degree structures to introduce specialty certificate and diploma programs in geospatial science.

This is reflected in the fact that in North America, of the 209 geography departments listed in the 2017–2018 AAG *Guide to Geography Programs in the Americas* (United States (185) and Canada (24)), 96% ($n = 200$) offered a GIS specialty qualification (United States (179), Canada (21)) and slightly more than half (57%) offered a GIS certificate program (United States (115), Canada (5)) (AAG 2018).[5] Further, in a recent publication that conducted an overview of Canadian institutions, 94 universities and colleges were reported to offer some form of program related to geospatial information studies (Natural Resources Canada 2015). In the mid-1980s, fewer than ten geography programs in the United States and Canada offered GIS courses, and none of these offered specialized degrees, diplomas, or certificates (Zhou et al. 1999). Hence, there has been considerable growth in the number of tertiary-level geospatial science programs available to prospective students.

With the rapid growth of such programs in higher education, and in the absence of any broadly accepted national standards to follow in Canada, the nature and content of curricula have become misaligned with the job preparation expectations of students. Two initiatives in the United States provide specific and explicit guidance on this problem by (re)aligning workplace skills for geospatial occupations with respect to academic curricula. Specifically, the Geographic Information Science and Technology (GIS&T) Body of Knowledge project provides a comprehensive

[5]The AAG publication includes institutions who voluntarily submit their information. Thus, the information may not be a complete reflection of all existing geospatial programs.

codification of more than 330 topics, organized into 73 units and 109 knowledge areas with 1660 learning objectives, that serve as a reference point for curriculum planners to prepare students appropriately for the workplace (DiBiase et al. 2006). Second, the US Department of Labor's Employment and Training Administration (DOLETA) introduced in 2010 a revised hierarchical Geospatial Technology Competency Model (GTCM) comprising technical and nontechnical competencies that characterize occupations within the geospatial sector (DiBiase et al. 2010).

Reinforcing the importance of the GIS&T Body of Knowledge and the DOLETA GTCM, multiple levels of government, retail and service-based businesses, and many forms of resource extraction industries in Canada are increasingly drawing on aspects of geographic knowledge as routine parts of their workflows. These aspects include, among others, spatial analysis, data integration and visualization, and reporting complex spatial data in map and other easily understood forms (Gedye et al. 2004; URISA 2007, 2011). A specialized degree in geospatial science provides students with these skills, and combined with more general spatial thinking and problem-solving associated with geography courses, graduates should, based on the demand noted earlier, be capable of finding employment in a wide range of careers (see Solem et al. 2013 for a list of geographer profiles). However, in practice, and contrary to the overall employment outlook of Employment and Social Development Canada,[6] there appears to be an unsettling imbalance in the *actual* number of opportunities relative to the number of higher education graduates seeking employment. The net result is that many students are left struggling to find work that is related to their education.

As an example of this trend, only close to half (45.9%) in a recent survey of Quebec university graduates of geography with a bachelor's degree had found a job, and only half of that subgroup worked in a position related to their area of study. The remaining students either returned to study (47.7%), became inactive (not seeking employment, not employed, and not studying) (3.5%), or were unemployed (6%) (Gouvernement du Québec 2013a). Masters graduates were slightly better off with 66% working full time, and, of that group, 89.7% were working in a related field. However, the same proportion of Masters graduates were unemployed (6%).

The success levels are somewhat higher for college students in specialized geospatial science programs. For example, within the Quebec college system in 2013, 80% of graduates with a geomatics specialization in cartography were employed, and of those all were employed in a field related to their course of study. However, unemployment among college graduates was slightly higher, and success levels for closely related "Land Use Planning and the Environment" courses of study in the college system were low. Only 48% of students were able to find work, and of those not all were working in a related field. The remainder were either continuing with a course of study, unemployed, or inactive (Gouvernement du Québec 2013b).

Students studying geodetic surveying within both the university and college systems had substantially higher success rates in both finding work in general and in a

---

[6] Government of Canada's Career Tool provides broad employment data of university graduates in geography: https://www.jobbank.gc.ca/studentdashboard/FOS20705/LOS07

related field. Moreover, when those continuing on to study are excluded, unemployment among graduates was zero. Hence, unlike degree or diploma programs such as geodesy (surveying) and geomatics-geodetic surveying, where the route to employment in a relatively well-recognized profession is more straightforward, geography and vaguely defined geospatial science-related jobs tend to be problematic for graduates. This is likely because many positions which require the skills and interdisciplinary knowledge of a geographer rarely include the word "geography," "geospatial science," "geographer," or "geomatics" explicitly in the job title or description. In fact, in Canada's largest provincial job market (Ontario), the keyword "geographer" failed to return any available positions using the provincial Ministry of Training, Colleges and Universities Web site "Find an Occupation"[7] tool (as of November 8, 2018). Nationally, of the 82,624 positions available, none used the term "geographer," 5 included "GIS," and 39 mentioned "geospatial." Clearly, while not all available positions are listed in the provincial and federal job banks, these results are not promising for graduates of geography or geospatial science programs.

Hence, over the last decade, geography graduates with geospatial science specializations have increasingly had to define their own career path, even when they have incrementally built related knowledge and technical skills across multiple courses at university or college. Compounding this issue, only a relatively small number of university programs involve aspects of explicit experiential learning, such as cooperative education and paid or unpaid internships that provide students with workplace experience as a part of the formal education process. Given that limited well-informed career guidance is available to students (Dietsche 2013), it is difficult for many to navigate the transition from education into the workforce. This point is returned to later in the chapter.

Despite this potentially worrying picture, some good, yet underutilized, general resources exist for students to help in their career planning. For example, the American Association of Geographers,[8] the Canadian Association of Geographers,[9] and the Royal Geographical Society[10] have produced a collection of online resources for career exploration. In addition, a practical overview and guide to geography professions can be found in Solem et al. (2013). The latter book brings together authors from academia, nonprofit, government, and industry, to present multiple navigational aids for students to assist their search for relevant employment.

The next section builds on the dilemma facing many students in geography and the geospatial sciences by reviewing the characteristics and profile of the global geospatial sector. This places the above discussion of employment prospects into perspective internationally as well as nationally within Canada.

---

[7] Ontario: https://www.iaccess.gov.on.ca/labourmarket/search.xhtml?lang=en
Federal: https://www.jobbank.gc.ca/home

[8] American Association of Geographers, Jobs and Career Center: http://www.aag.org/careers

[9] Canadian Association of Geographers, Profiles of Professional Geographers:
https://www.cag-acg.ca/profiles-of-geographers

[10] Royal Geographical Society, Careers resources:
https://www.rgs.org/geography/studying-geography-and-careers/careers/

The Geospatial Industry and Employment Prospects for Canadian Students

The global geospatial industry can be described in its simplest form as having three cornerstones, namely, education, government, and the private or business sector. Together these sectors comprise the communities of knowledge workers, data and software suppliers, and consumers. The latter two sectors have spatially variable needs for skilled workers across multiple diverse domains, and the former sector meets the need for knowledge workers through the supply of graduates from university- and college-based geospatial education programs. Despite the relative simplicity of this three-cornered model, there is as much, and possibly even more, complexity within each of the sectors as there is between them. This complexity compromises the ability to make comparisons and limits discussion necessarily to high-level observations. Nonetheless, some important general trends can be discerned.

Producers and consumers of spatial data and software can be characterized and related to the prospects for graduates in terms of employment opportunities. Numerous global industry reports are produced annually. However, there is little consistency in their estimates of global market value, the market value for individual countries, or predictions for the future. For example, according to Daratech, a global market research firm, between 2009 and 2010, the geospatial industry grew worldwide by 10.3% to reach a value of US$4.4 billion, with a forecast of an additional 8.3% growth to almost US$5 billion in 2011.[11] A more recent report published in January 2012 by a similar company, Global Industry Analysts, suggests that the GIS industry was expected to grow by this year (2015) to a worldwide market value of more than double that suggested by Daratech (US$10.6 billion)[8]. However, these estimates pale relative to figures provided through detailed case studies undertaken in specific countries.

For example, in Australia, ACIL Tasman reported in 2006–2007 the impact of the spatial information industry on the Australian economy alone as ranging between AUS$6.43 and AUS$12.7 billion (ACIL Tasman 2008). An equivalent study conducted by the same company the following year in New Zealand, which is a significant minnow compared to the value of geospatial data to the North American economies, estimated the impact of the geospatial information industry as adding NZ$1.2 billion to the national economy. This contribution was considered to be a result of the increasing adoption of modern spatial information technologies over the previous 13 years, and it equated to 0.6% of New Zealand's GDP in 2008. Other nonproductivity benefits linked to the increasing use of spatial information were thought to be worth a multiple of the $1.2 billion contribution (ACIL Tasman 2009). Studies in the United Kingdom (England and Wales) in 2010 by ACIL Tasman and ConsultingWhere, and by Indecon International Economic Consultants in Ireland in 2014, report similarly significant value added by the geospatial information industry to the respective economies.

---

[11] GIS Lounge: GIS Industry Trends and Outlook, http://www.gislounge.com/gis-industry-trends/

Following the Australian and New Zealand market research, a similar study was conducted in Canada in 2015 by Hickling Arthurs Low Corporation in conjunction with ACIL Allan Consulting, Fujitsu Canada, and ConsultingWhere, under the general management of the GeoConnections program of Natural Resources Canada's (NRCan) Mapping and Information Branch. This study reported that in 2013 the Canadian geomatics industry was comprised of 2450 predominantly small firms with less than 50 employees but that together they contributed Can$2.3 billion to the national gross domestic product (GDP), with geospatial information contributing Can$21 billion toward GDP. The report also suggested that geospatial information generates 19,600 full-time equivalent jobs across all sectors of the economy. Despite some differences in the numbers provided throughout the report, the impact of the industry can be said to be profound. However, while prospects for continued growth of the sector are described as "promising," with the impacts of geospatial information described as having "enormous potential," there is no tangible evidence or real discussion of the opportunities the sector offers for the swelling ranks of university and college graduates exposed to some aspect of geospatial technologies during their postsecondary education.

Hence, in support of Stigler (1961), the economic impacts of information-based industries on national economies and collectively on the global economy are substantial. All available predictions from the ACIL-group studies noted above suggest continued growth, as more and more sectors of the economy discover the power of spatial information and the importance of location in decision-making. However, the transformation of this promise into tangible job opportunities is far less clear-cut.

Under normal circumstances, the analyses of the economics of the geospatial industry noted above would be good reason for optimism on the part of Canadian students of geospatial science. However, these have not been normal times of late, especially in Canada. Linnitt (2013)[12] and especially Turner (2013) in his well-researched book *The War on Science* suggest a less optimistic future for Canadian graduates than the numbers provided in the ACIL-group 2015 report might suggest. In contrast, Linnitt and Turner provide details on how the previous federal conservative government of Stephen Harper systematically, over its two terms in office, reduced the government's capacity to gather data and downsized or eliminated offices that collect, monitor, and analyze scientific information.[13] In fact, the Harper conservative government seized control of the channels through which science is communicated, prevented the publication of research that might interfere with the agenda of resource-based private industry, and promoted rapid resource extraction by dismantling an entire century's worth of environmental regulations, environmental monitoring, and data-based science.[14]

---

[12] http://www.academicmatters.ca/2013/05/harpers-attack-on-science-no-science-no-evidence-no-truth-no-democracy/

[13] See "When Harper Killed the Census He Robbed Canadians": http://www.huffingtonpost.ca/murtaza-haider/harper-long-form-census_b_5614355.html

[14] See review by Travis Lupick: http://www.straight.com/life/501106/new-book-chris-turner-lays-bare-stephen-harpers-stifling-war-science

Indeed, restructuring at NRCan, the federal mapping agency, in 2012 eliminated 100 positions, and the agency expected its operating budget to fall from $3.5 billion in 2012 to $2.4 billion in 2013–2014. At the same time, the number of permanent, full-time positions decreased from 4389 to 4155.[15] As a result of these cost-cutting measures, morale is not high in the sectors traditionally aligned with geospatial data collection, its dissemination, and use within the Canadian economy. The election of a new liberal government in Canada, with an agenda to invest in new growth rather than cutting costs, and on openness rather than closure in consultation and decision-making, may serve to provide better prospects for new graduates.

However, further affecting the prospects for the geospatial sector in the coming years, the recent downturn in the global economy as a result of dropping or stagnant crude oil prices has hit Canada's resource-driven economy, and its center of operations in the province of Alberta, especially hard. Where, then, does this leave the Canadian university and college graduating classes and the starting cohorts of 2018 through to 2022? Statistical evidence suggests that the future is not necessarily that bright.

For example, using Statistics Canada data, Coates and Morrison (2012) report that young Canadians will continue to bear the brunt of a continuing weak job market. The number of employed 15–24-year-olds fell by 300,000 jobs between 2008 and 2012, and the unemployment rate for this group in 2012 was 14.8% (in January 2015, 12.8% compared to 5.5% for those aged 25 and over). Of more significance, Coates and Morrison (2012) also note that in 2010 the level of *underemployment* (university graduates working in positions that require no postsecondary qualifications)[16] was more than 33%, second only to Spain in the Organization for Economic Co-operation and Development (OECD) countries.

On the other hand, the Association of Universities and Colleges of Canada (AUCC 2011) has predicted that there will be close to 1.3 million more jobs for university graduates in 2020 than there were in 2010, and to meet this demand, the number of new enrolments and graduates will have to increase by 1.3% per year over the course of the next decade. However, these are total jobs, and, given the state of the Canadian economy noted above, it is unlikely that more than a relatively small number of these new positions will focus on jobs that require the technical skills and body of knowledge that geography and geospatial science graduates acquire.

This will not stop young high school graduates from seeking a career built on tertiary-level study. Compared to other OECD countries, Canada has one of the highest rates in the world of continuity from high school to higher education within 2 to 4 years of high school graduation. This is unlikely to change in the foreseeable future. For example, in 2009–2010, 75% of all graduating students from high school continued on to college or university, and most of these individuals completed their

[15] http://www.canadaforchange.ca/2012/01/25/restructuring-at-natural-resources-canada-to-eliminate-100-jobs/

[16] CBC produced a documentary in 2015 on unemployment among university graduates, 'Generation jobless'

chosen course of study. These numbers elevated Canada half a decade ago to one of the highest rates of postsecondary graduation (over 50%) qualifications among 25–64-year olds in the world (OECD 2013).

The collision of these converging factors suggests that current university and college undergraduate geospatial science curricula must be crafted with great care, certainly relative to the era when specialized curricula first appeared in this field. Rather than rushing to graft on new programs to attract additional students, focus must equally be given to ensuring that current programs are imparting knowledge that is relevant to the information needs of the modern economy. As suggested earlier in this chapter, the era of knowing how to use a mainstream GIS software platform to create, process, and use spatial data is now no longer enough for graduates to gain direct employment in related fields.

Specialist students must be afforded the opportunity to go beyond basic knowledge and to learn details of spatial data and databases both for server-based distributed multiuser networks and, especially, Web- and cloud-based spatial information architectures. Moreover, to enhance employment opportunities, Web mapping and geospatial application development, which in the past was "nice to know," are now a matter of "must know" and more than likely a matter of "must know deeply." The same can be said for mobile GIS, 3D GIS, LiDAR, and other areas, which in the past were considered specialist, but are now becoming integral to the mainstream geospatial information industry. In addition to the required technical skills noted above, attention must also be given to enhancing soft skills, perhaps understood as personal attributes (Borwein 2014b) or non-cognitive skills (CCCE 2014b), of prospective employees. These are explored further in section "Key Competencies for Successful Performance in the Geospatial Workplace".

Despite the troubling job statistics relative to the number of graduates currently exiting geospatial degree and certificate/diploma programs, the overall move in Canada to a digital economy may hold some promise for future employment prospects. Federally in Canada, this strategy was supported by a review in 2011 of research and development to determine how it can encourage private-sector innovation to help drive higher productivity.[17] The net outcome of these business-driven strategies is likely to push even higher the labor-market demand for highly skilled graduates, and this has clear implications for students well versed in interdisciplinary and specialized aspects of geospatial data use. However, growth, if it does occur, is likely to fuel all sectors of the economy, and this may also serve to drive incomes upward for less-skilled jobs and thereby increase the opportunity costs of attending university rather than following a path into a skilled trade directly from high school. Hence, lower growth scenarios seem more likely in terms of the overall number of university and college graduates as well as jobs, as a flow through from demographic transition-driven lower enrolments and the effects of policy and economic restructuring take effect.

To be truly competitive in the job market, in addition to technical skill-based learning and new knowledge acquisition, students must learn workplace skills

[17] http://rd-review.ca/eic/site/033.nsf/eng/home

such as how to plan strategically and meet goals, communicate with colleagues, and take leadership (Schlemper et al. 2014). Too often these important dimensions of knowledge are relegated in favor of increased focus on technology use. Simply stated, while deep technical knowledge is better than basic knowledge, it is still not enough for an individual to thrive in the workplace. Graduates must know how to formulate realistic goals, achieve objectives within workflows, and craft strategic task-based approaches that can achieve success. The ability to do this well is at least as important as technical knowledge, and it should figure prominently in all undergraduate curricula.

In general, it is reasonable to suggest that a tipping point is fast approaching where university- and college-based geospatial programs are going to enter a new phase of curriculum design. The first phase in this process comprised courses that involved various levels of GIS instruction as part of a conventional undergraduate geography degree. The second phase that started in the early 2000s saw the introduction of specialist certificates of accomplishment or diplomas in GIS in addition to a geography or related degree. This was followed by a sub-phase in the mid-to-late 2000s that produced specialist geospatial degree programs. The third and current phase is characterized by widespread and increased availability of specialized certificates, diplomas, and stand-alone degrees in geomatics/GIS. Unfortunately, these programs have grown for the most part without any close attention to the content of either the DOLETA GTCM or the GIS&T Body of Knowledge discussed earlier.

The next differentiator in higher education will likely be a phase of deeper specialization within specialist degrees, perhaps with certification in Web GIS, mobile GIS, or 3D GIS, accompanied by paid cooperative education or unpaid internship experience that equates to a year of full-time work experience upon graduation. This experience, combined with greater focus on the soft skills noted above, may help to produce much more rounded and therefore readily employable graduates in future years.

However, not all schools can implement this level of program readiness, and even those with such programs may still struggle to launch their students into appropriate job placements. Those that can achieve this level of program content will emerge as the most competitive in a highly competitive education marketplace, where students can vote with their feet by choosing the institution that offers them the best educational experience *plus* the smoothest and best verified transition into a job. The next section discusses the key competencies that are valued by employers, and that forward looking geospatial science programs should seek to cultivate.

Key Competencies for Successful Performance in the Geospatial Workplace

As suggested above, the geospatial workplace is diverse, with potential career opportunities in many fields. Its amorphous nature makes it difficult to define with any real precision what geospatial professionals do and what skills they possess or require (DiBiase et al. 2010). On the one hand, this diversity is a strength in that it allows graduates to explore opportunities across multiple sectors. On the other hand, it is also a weakness as the subject area continues to lack widespread recognition in job titles and descriptions, and the breadth of potential options makes it difficult to navigate the field and build competencies that align with employers' needs.

It is clear that to be competitive in the modern workplace, graduates must develop competencies that include not only deep domain knowledge and skills but also other abilities that are required for high-level tasks (Gaudet et al. 2003; DiBiase et al. 2010). In addition, students must also focus on building the soft skills that are important to succeed in most workplaces (McKendry et al. 2014; Schlemper et al. 2014). These hard and soft competencies figure prominently in the only broad-based model that has been developed for the geospatial sector, namely, the current DOLETA GTCM noted in the section "Introduction". Although this model, and its earlier forerunner, was developed for the US employment market, there is sufficient similarity with the Canadian geospatial market to use it as a benchmark to tease out the competencies that Canadian educators must include in curricula and that students should seek to build during their academic career.

The first GTCM was developed by Gaudet et al. (2003) well over a decade ago. It outlined four general areas that included specific technical (13 total with 4 core), business (11 total with 4 core), analytical (6 total with 2 core), and interpersonal (9 total with 5 core) competencies. The core competencies in each of the 4 classes were associated with 12 specific work roles that described tasks in the general geospatial sector. Together, the 39 total competencies included multiple technical and nontechnical areas of knowledge and abilities that were important for workplace success (Gaudet et al. 2003). This initial model was an important contribution to creating oversight of the geospatial technology workplace and its requirements for knowledge workers. However, and perhaps inevitably, it came under criticism within half a decade on the grounds that its 13 technical competencies did not fully reflect the requirements of a rapidly evolving field (DiBiase et al. 2010).

Hence, the initial GTCM was revised incrementally, with the latest version spearheaded by the GeoTech Center and published during the course of 2013–2014.[18] The process of revision, as with the original GTCM, used significant input from industry experts to produce a pyramidal model, with each of its nine tiers representing distinct competencies in the geospatial field (see DiBiase et al. 2010 for a comprehensive account of its development). The arrangement from bottom to top progresses from broad-based personal effectiveness competencies to specific industry competencies. The foundation competencies at Tier 1 include attributes such as interpersonal skills, professionalism, and initiative. Academic competencies across a range of related subjects including mathematics, geography, basic computer skills, writing, and critical and analytical thinking are found at Tier 2. Tier 3 focuses on workplace competencies including teamwork, planning and organizing, problem-solving, and decision-making, among others. Tier 4 specifies a range of

[18] The current DOLETA GTCM can be found at http://www.careeronestop.org/competencymodel/competency-models/geospatial-technology.aspx

core geospatial abilities including 43 specific core components of knowledge and skills. Tier 5 focuses on sector-specific technical competencies such as analysis and modelling, software and application development, and data acquisition. The final tier at the peak of the pyramid includes occupation-specific requirements and management competencies.

Both the original GTCM and the DOLETA GTCM emphasize the importance of acquiring, through education and training and/or through on-the-job activities, technical or hard skills and nontechnical or soft skills by rounded employees. However, formal geospatial education programs both in the United States and in Canada have tended to focus on the technical aspects of geospatial science, with relatively little explicit attention given to the development of well-rounded skills. The relative underrepresentation of intellectual knowledge (e.g., problem-solving, explanation), personal attribute skills and knowledge (e.g., time management, collaborative group work, writing and presentation skills, knowing and understanding the geospatial industry, work experience), and university or college life elements (e.g., networking) remains absent from most tertiary geospatial programs in Canada.

The importance of having a blend of technical and nontechnical competencies is also evident in a series of surveys conducted first in 1998, next in 2003, and then every 4 years by the Urban and Regional Information Systems Association (URISA). The reports from these surveys include feedback from GIS/IT workers primarily in the United States but also in Canada and internationally. Similarly to the data from Quebec presented in section "Introduction", the majority of respondents in the most recently reported URISA survey (2017) were employed by the government (57.7%), with 14.6% in the private sector (URISA 2017). Daily work described by the respondents included collaborating with colleagues from different departments, performing geospatial tasks, managing staff, as well as a range of analysis, writing, and presentation tasks (URISA 2011, 2017).

The URISA survey results recognize several types of hard, computer-based, skills that are important in the geospatial workplace including GIS and database management software use and programming in various languages (Visual Basic, Python, HTML, JavaScript, XML, and Flex, Silverlight). To put this information into better context, Johnson (2010) examined the specific competencies for eight entry-level GIS positions in the United States, which in principle apply equally to Canada. In total, 476 individual job tasks were identified, which were then categorized into 55 common job task categories, 35 knowledge and skill categories, and 27 behavior categories. This list was further distilled into eight "duty categories" including manage, generate, process, and analyze data, manage software, manage projects, generate products, and professional development (Johnson 2010). The knowledge and skill categories were weighted heavily toward hard skills, although importantly they also included communication (verbal, presentation, and writing), critical thinking/problem-solving, spatial thinking, organization, and time management skills. In contrast, the behavior categories were almost exclusively soft skill-oriented, and most were rated of high median importance by the "Developing a Curriculum" (DACUM) panel of experts used in the analysis.

Table 1 Description of soft skills in geospatial science roles (URISA 2017)

Job title	Soft skills mentioned in job description
Director of GIS/geographic information officer	"All GIS personnel… fall under the supervision and direction of GIS"
GIS specialist	"Often in a team environment, provides customer and technical support"
GIS coordinator	"Provides technical support to other agencies, individuals, and governments"
GIS manager	"Coordinate GIS activities between different groups"
GIS programmer	"Provide technical support to other GIS professionals"

While computer-based skills are relatively easy to teach through the learning software interfaces and procedures, soft skills such as those noted above are much more difficult to teach and must be approached using innovative approaches to curriculum content and teaching style. Simply stated, the "learn software from a workbook or textbook" approach is no longer as relevant to higher education as it may have been two to three decades ago.

While some professional geospatial roles clearly require soft skills (e.g., educator/trainer, GIS business development/sales and marketing, independent consultant), the same is true for more technically oriented roles. Table 1 summarizes five job descriptions, regardless of level, and explicitly includes the use of soft skills, especially interdepartmental collaboration. Even in roles that are technical in nature, soft skills are inherent to any workplace. For example, the 2017 URISA responses describe a GIS analyst as an individual focused on data and programming. However, for any given project, soft skills are needed to produce the desired program or analyses through discussion, clarification, and collaboration.

Comfort in teamwork is an essential soft skill to acquire as in most workplaces GIS staff do not work in an isolated environment. The tasks, issues, and problems within most organizations are far too large and complex for a single person to resolve. Hence, related soft skills that facilitate activities such as collaboration and professional conduct are extremely important. Their importance is evident in the results of a survey by member companies of the Canadian Council of Chief Executives (CCCE 2014b). Respondents valued soft skills more than hard skills for entry-level positions, paralleling findings reported in a national study (Smith and Lam 2013), as well as one from Ontario (Borwein 2014a; Refling and Borwein 2014). Hence, curricula in higher education should provide students with opportunities to develop these soft skills. However, conventional technical approaches to geospatial learning do not always bring these to the fore. This comprises a significant learning gap that may serve to compromise the ability of students to navigate relevant dimensions of the job market upon graduation.

A human resources manager[19] from an international nonprofit organization has the following perspective on the balance that is needed between soft and hard skills in the workplace, many of which are aligned to the different tiers of the DOLETA GTCM:

[19] This contribution is from a verified source who prefers not to be identified.

> As organizations have become more effective in their recruitment practices and in many cases, the competition for positions have become fierce. It is not just having a 4.0 GPA that will guarantee your foot in the door. In my experience, hiring staff with backgrounds in geography and GIS will always entail testing and assessments of the technical abilities of the candidate. But in many recruitment activities, we also seek competencies that are aligned to the organizational culture and from which we can estimate a candidate's expectation to succeed.

Below is a list of some of those soft skills commonly required in new hires:

- Communication: Speaks and writes clearly and effectively, listens to others, inquires appropriately and inclusively
- Technological Awareness: Stays abreast of new technology, constantly aware of and learning new technologies
- Teamwork: Collaborative and motivating, sincerely values others' ideas
- Planning and Organizing: Identifies priority activities and assignments; adjusts priorities as required, has project management skills (time, budget, resources), foresees risks, delivers in a timely fashion
- Accountability: Demonstrates ownership, supports more junior staff, responsibility for his/her own as well as team shortcomings
- Client Orientation: Identifies clients' needs, provides appropriate solutions, delivers regular progress reports, meets deadlines in a timely fashion
- Creativity: Outside of the box thinker, creative, and innovative
- Commitment to Continuous Learning

One means to foster these soft skills, while continuing to hone hard skills is through collaborative or group project work that mimics a real world work scenario. Consider, for example, the following group assignment:

> Your final group project is to take on the role of GIS consultants for a client who owns a multibranch bulk food grocery business (such as a supermarket chain). The client wishes to expand the grocery chain by adding two new store locations in a major Canadian metropolitan center. For this project you will work in a team of four. Your first task is to identify the current store locations owned by the client and all competitor locations. Next you must develop a strategy, collect information on the client's current market share, and store catchments relative to the share and catchments of the competitors. You will build this information into a spatial database that will allow you to undertake a comprehensive analysis of current and future market opportunities. Each team member must bring his/her own ideas and perspectives, perhaps based on personal knowledge or work-term experience to solve the problem. With multiple concurrent tasks to be completed, each member must take on responsibilities and meet deadlines, and the team must meet regularly to assess progress and problems. A final written report that contains a store expansion plan, supported by maps and data analysis, and recommendations that will improve the client's market share must be professionally prepared. The group must also present their recommendations verbally to the client.

Clearly, technical skills in spatial data collection, spatial analysis, and visual communication are required to solve this problem satisfactorily. However, it is equally valid to ask students to ponder individually in a separate report, upon the project's completion, the following questions:

1. List the soft skills you have used to complete this project.
2. What were the top three soft skills you used in the work and why?
3. What previous experiences (e.g., in coursework, work terms, previous experience, etc.) did you use to hone the soft skills you used in this assignment?
4. Which soft skills did you find most challenging during the project?
5. How could future course work help you to improve the use of these skills?

Soft skills can be learned through many means including formal education, volunteer roles, related or general work, and other community or leadership activities. For college and university instructors, it is often instructive, especially in smaller third or fourth year classes, to ask students to enumerate the exposure they have had to these types of skills. This can be easily achieved by asking students at the start of a course to identify and complete in the right-hand columns of Table 2 descriptions of the soft and the hard skills they feel they already have. The table is organized into three sections, including formal education, work experience, and community or leadership activities, where students can list all of the skills that they feel they have acquired to this point in their career. A key goal of this exercise is to help students recognize their strengths and areas that require growth so that they take the initiative to develop the skills to manage (e.g., lead and delegate group project tasks), analyze (be familiar with new analysis tools), present findings (give presentations, dissemination on relevant social media platforms), and engage in leadership (initiate solutions) roles, which are all valuable skills to prepare individuals for employment (McKendry et al. 2014; Schlemper et al. 2014; Monk et al. 2012).

At the end of the course, the same exercise could be repeated, noting any changes that the course has helped to nurture. For ideas of relevant skills, students should be encouraged to consult the DOLETA GTCM online and click on the competencies that characterize each tier. In addition to providing guidance on course content and

Table 2 Current hard and soft skills template developed through formal and informal experiences

Education, work, community activities	Current soft skills known	Current hard skills known
Formal education		
College/undergraduate degree		
Graduate degree/continuing education		
Certificate program		
Work experience (paid or unpaid)		
Internship, coop, work study		
Related work experience		
General work experience (e.g., administrative staff, server)		
Personal projects		
Community and leadership activities		
Volunteer		
Student leadership positions (e.g., leader of student group or club)		

approach for its next session, this process can have the added advantage of providing students with a useful list of skills linked to an industry standard competency model that they can include in their resume and explain in a cover letter.

A simple exercise such as this can be supplemented by asking students to compare their own soft skill competencies with those that are required in the criteria of an entry-level geospatial position. This activity follows the approach outlined in Ferguson's (2014) article that poses the question of "what can I do with my degree?". This question should be asked and answered multiple times by students during their college or university career. Consider the following job requirements derived from the GoGeomatics Canada Web site job board on November 8, 2018. Students should be encouraged to read the description and answer the questions that follow.

Position: GIS Analyst[20]

Responsibilities:

- Preparing and maintaining documents, presentations, displays, graphics, 3D models, and maps using analytical and design software for print and digital distribution
- Gathering, compiling, migrating, and converting datasets from a variety of internal and external sources
- Digitizing land surface features, such as river centerlines and banks
- Building digital elevation models by integrating topographic and bathymetric data
- Acquiring, rectifying, and mosaicking aerial imagery products

Qualifications:

- A university degree in GIS, cartography, geography, civil engineering, or related field of study
- 2+ years of GIS experience
- Educational or professional experience with ESRI ArcGIS and its extensions
- Experience with client-server GIS technologies including ArcGIS API for JavaScript, Flex, and Web GIS/Web Mapping (i.e., MapOptix or Geocortex Essentials) platforms is desired
- Ability to manage multiple tasks and deadlines
- Demonstrates a high degree of attention to detail and pride in work products
- Ability to build and maintain positive working relationships with co-workers and clients

[20] This criteria list is a merge of two job postings found on GoGeomatics Canada on November 8, 2018 (http://www.gogeomatics.ca). It is highly recommended that Canadian geospatial science students visit this and other job boards regularly during their undergraduate careers to ensure that their knowledge is relevant based on the current needs of employers in the geospatial sector. This provides an excellent instructional direction for course selection and for the development of both hard and soft skills.

After reading the job requirements, answer the following self-assessment questions:

1. What experiences have provided you with the hard and soft skills that fit either the responsibilities or qualifications sections?
2. What qualifications might you need in order to upskill or learn?
3. How might you expand the hard and soft skills that are evident in the job requirements?

There are numerous ways that students can build additional competencies both within their regular course of study in a geography and geospatial science degree or diploma, by taking additional online courses such as Esri Virtual Campus.[21] In addition, there is now a growing range of Web-based massive open online courses (MOOCs) that focuses on aspects of geospatial skills. In addition to MOOCs, there are many other Web resources also accessible for free upskilling, not only for students but also for those already in the geospatial workforce. For example, anyone with an Internet connection can gain experience with the Esri JavaScript API by accessing the Esri Canada lesson planner Web site[22] and working through a freely available learning framework for the JavaScript API that includes the use of stylesheets and the dojo development toolkit.[23] Completion of this learning framework will allow students to learn how to make a server-side, customized Web map service (see bullet four in the "Qualifications" job posting noted above).

While such resources are available to enhance existing hard skills, the same is not true to the same extent for the soft skills discussed above. However, the importance of soft skills that can help to lead to a successful job search cannot be underscored enough. The next section explores ways to develop and broaden students' soft skills through various activities.

Enhancing Soft Skills for Improved Employment Prospects Among Geospatial Science Graduates

Statistics from 2012 to 2013 in a survey of Canadian employers ($n = 920$) indicate a fall on average from 16 to 9 new graduate hires (Smith and Lam 2013). The same survey shows a clear preference for hiring students with a good blend of hard and soft skills and who worked at the same company through internship, cooperative education placements, or work-study programs. This may be driven by the assurance that students with prior work experience at a company will understand the workplace culture and can contribute seamlessly without the need to learn new workflows (CCCE 2014b).

---

[21] https://www.esri.com/training/

[22] http://hed.esri.ca/resourcefinder and use the keyword "javascript"

[23] http://dojotoolkit.org/

This emphasis on career readiness upon leaving higher education is responsible, in part, for the changing nature especially of university programs. Up until relatively recently, university-based education was considered distinct from college (vocational) courses by focusing on the intellectual knowledge growth of students rather than practical skills that were typically more associated with training. Hence, the path from university to work was not explicitly considered, and it was assumed that the job market would essentially absorb new, well-educated graduates into various career paths. However, the growth, noted earlier, of technical skill-based undergraduate and graduate diploma and certificate programs in subjects such as geospatial science has blurred the traditional separation of education, training, and work. As a result, many colleges have become degree granting without substantially changing their emphasis on training, and many universities have introduced explicit practical workplace preparation components in their courses of study.

For example, upon its foundation in 1957, the University of Waterloo (UW) was a Canadian pioneer in cooperative education for its engineering students (Tamburri 2014). Over the past 50 plus years, UW has grown its cooperative education program to become the largest in the world, with approximately two-thirds of all students enrolled in over 140 accredited programs. Students value the importance of this work placement approach during their undergraduate careers relative to their improved employability upon graduation (Tamburri 2014). A former geospatial cooperative graduate of UW describes her entry into the job market as follows:

Janice Lee, Financial sector, Canada

When I chose to focus my university studies on geography and, in particular, geomatics, I knew it was important to take the cooperative education option. Although taking the cooperative education would mean graduating 8 months later than a regular stream of studies, I felt it would be valuable to experience different geomatics jobs in preparation for a career after graduation. I was fortunate enough to work in various departments in the government and see how GIS and remote sensing were applied – from capital works planning to map publishing, GIS software testing, and academic research. These opportunities certainly fostered and taught me new technical skills, as I had hoped. Just as important, these experiences further developed my nontechnical skills – working in a team, interpreting people and situations, communicating ideas and deliverables. The classroom lessons on GIS, from spatial thinking to software application to project proposal writing, built a good foundation for a career in the GIS industry.

To enhance traditional lecture and lab-based learning approaches while enhancing their market appeal and lifting student enrolments, an increasing number of universities and colleges, not just in North America but throughout the world, are adopting work placements as part of their degree programs. Everyone benefits from this mixed-mode approach to learning, providing that both hard and soft skill development are kept central to the learning process. Students benefit in the sense that they have the opportunity to learn workplace skills while studying, while employers can also pre-screen potential employees, give experiential opportunities to the education community, add specific skills or talents to their organization, and support special projects or relieve short-term staffing pressures (Sattler and Peters 2012). Students also benefit from making themselves more competitive in the highly competitive and tightly constrained current job market.

The University of Regina, also with a long tradition of integrating cooperative education into the curriculum, developed the UR Guarantee Program[24] to guide students' career exploration. In this program, undergraduate students receive career-related support throughout their degree leading on to successful employment. Students who do not have employment within 6 months of graduation are eligible to take another year of undergraduate courses without charge. Other Canadian universities are taking similar actions to encourage student involvement to develop soft skills. Some universities have initiated a cocurricular record. Similar to a transcript, this is an official institutional record that documents students' involvement in eligible student activities. The idea is that institutions recognize students' learning outside the classroom and have documented evidence to show their future employer.

Whereas hard skills can be improved through repetitive or rote learning and studying, soft skills require development through interacting with people and experience. Efforts to enhance students' professional attributes are also noted at the graduate level. For example, some institutions host workshop events, free to current students, ranging in topic from leadership, language training, research management, strategic communication to wellness and life balance.[25] Hence, any type of "learning by doing" can be helpful in achieving the three following aspects of a rounded education.

Relevant and General Work Experience Work experiences give value to students because they provide a lens through which to establish work preferences (e.g., aversion to desk-bound tasks such as online technical support), strengths (e.g., coding to customize aspects of a GIS project), and desired work environment (e.g., work in a collaborative team within an open office). By experiencing various types of work and workplaces during a course of study, students give themselves the opportunity to reflect on their own strengths and weaknesses while applying the knowledge they have gained from formal learning. They can also develop transferable skills that are important in terms of future career prospects (Helyer and Lee 2014). Similarly, employers value evidential work experience shown by potential job candidates (e.g., work-terms, internships, and service-oriented learning), as well as extracurricular activities (e.g., clubs, societies, sports) in association with academic performance (Sattler and Peters 2012; Smith and Lam 2013; CCCE 2014b).

Community and Leadership Activities (Volunteering) At any stage of their education, students can explore opportunities to advance their interpersonal competencies by becoming involved in professional organizations such as URISA or the Canadian Association of Geographers, joining local community groups, or taking

---

[24] UR Guarantee Program: http://www.uregina.ca/urguarantee/about/policy.html

[25] For example, Concordia University offers a large selection of professional development workshops for undergraduate (FutureReady) and graduate (GradProSkills) students.

on leadership roles within university or college. Participation in such activities opens up new connections while fostering the informal development of important soft skills (e.g., communication, working with others in a team, and management).

By taking on leadership roles in university- or college-based clubs or societies, students can build key skills that are important in the workplace (e.g., being innovative or facilitating an event). As noted above, employers recognize the value of these activities and look favorably on students who can include these experiences in their resume (HECSU/AGCAS 2010; Smith and Lam 2013; CCCE 2014b). Whereas higher-level technical skills are definitely advantageous in terms of employability, they are also prone to rapid change, and frequent upskilling is necessary to stay current. In contrast, the softer and less-transient skills obtained by individuals skilled in networking and leadership are also important attributes (Helyer and Lee 2014).

The importance of community and leadership activities, as well as building connections, is clear in the following comments:

Lynn Moorman, Professor, Mount Royal University, Calgary, AB

If everybody in the pile of "credentials met" has strong geospatial skills, the decision on who to hire is then based on who is a best fit for the company or institution. This decision is based on interests, on enthusiasm, on how you have applied your knowledge (e.g., volunteering, community projects, teaching), and on what you can bring to the employer to enhance the workplace. The beauty of the geospatial world is that any life experience on the planet increases your spatial experience and can be relevant to your job, including travels, interests like hiking or skiing, and any type of field experience. My first GIS job with Canada Post was sealed when the interviewer saw I had spent summers on the Mackenzie River on a zodiac boat doing field work. My research associate position at the Canada Centre for Remote Sensing was secured when the interviewers saw I had a good knowledge of biogeography across the country from travelling and field work. And all along, the network of people from school (undergraduate and graduate) and early jobs have been critical in opening up opportunities, providing references, supporting projects, and building networks. People are bridges to opportunity. Treat colleagues at school and work well and with respect, and maintain friendships and connections now, because they can be valuable allies in your future, providing you with numerous ways to help you succeed.

Networking In accordance with the above comment, perhaps the best advice that can be given to students early on in their university or college career is to start building a network of contacts not only among their classmates but especially within the field or area they intend to work in. Creating, developing, and maintaining relationships, personally and professionally, is one of the most important activities in an individual's work or career (Forret and Dougherty 2001). Building an employment-related network can have many important implications in terms of shaping and directing future career prospects (Friar and Eddleston 2007; Wolff and Moser 2009). In fact, Koss-Feder (1999) estimates that 70–80% of the best jobs are a direct result of networking rather than simply responding to an advertised position.

Table 3 Personal network construction template

Education, work, community activities	Current network member names	Potential network member names
Formal education contacts		
Within college/undergraduate program contacts/roles		
Graduate degree/continuing education program contacts/roles		
Certificate/diploma program contacts/roles		
Workplace contacts (current and potential)		
Internship, coop education, workplace study contacts/roles		
Potential workplace contacts/roles		
Personal project contacts/roles		
Community and leadership activity contacts (current and potential)		
Volunteer contacts/roles		
Student leadership positions (e.g., leader of student group or club)/roles		
Other activity contacts (current and potential)		
Professional organization contacts/roles		
Information interview (cold-call)		
Other		

A recent survey found that graduating students have sought career advice from their network of friends (78%), parents (77%), or professors (51%) (Canadian University Survey Consortium 2015). Hence, a useful exercise to conduct with students is to have them list and assess their career-related network using Table 3 as a blank template.

This activity will not only propel students earlier rather than later into understanding the value of constructing a career network, but it may also assist them in adding names they may not have previously considered as important nodes to build job contacts around. These contacts, in turn, may become potential sources of information about opportunities for work-term employment, internships, or summer employment. In addition, organizations or individuals not currently part of an individual's network can be identified as potential contacts who can enhance future opportunities.

Students should be encouraged to think of all aspects of their course of study as a resource and to ask questions such as "what employment advice can you give me?", "what were your successful job search strategies?", "what was your work trajectory?", and so on. Very few individuals working within the geospatial industry would be unprepared to answer such questions, and assembling an inventory of answers can form an invaluable resource that can only help during the job search process.

Hence, gaining work experience during academic study, taking on leadership and extramural professional society and community roles, and building and using a job-related network are all essential aspects of being able to navigate successfully the search for employment prior to graduation.

Conclusion

This chapter has taken a wide-angled view of the Canadian geospatial science sector and its current ability to absorb the large, and constantly increasing, number of new graduates emerging from university and college programs. Views differ on the robustness of the sector, but what is clear is that geospatial science is a significant contributor to national economies in all economically advanced countries across the globe. Relative to the 1990s, and even into the mid-2000s, the marketability of a qualification in geography or a related field with a specialization in GIS is no longer an entry ticket to the geospatial workforce. The large number of college and university specialist degrees, certificates, and diplomas that now abound throughout the Canadian higher education system means that additional differentiators must be attained to stand out from the crowd.

Discussion in the chapter suggests that the students who will be most successful in navigating through the various channels in the voyage from studying to working will be those who enter programs that have at least three important characteristics:

1. The opportunity to learn, in the sense of deep learning, *geospatial technologies* currently used in the workplace.
2. Allow students to achieve *practical workplace experience* during their course of study through well-organized and paid work-terms, unpaid internships, or some combination of the two.
3. Expose students not only to *deep technology understanding*, but also promote intellectual knowledge building (i.e., problem-solving and spatial thinking) *built* around the *soft skills* that are crucial for workplace success.

Of course students are not simply passive receptors of learning, and they too can avail themselves of numerous points of advantage by being proactive during the course of their studies. These attributes include activities such as starting the construction of a work-related network as soon as possible. In addition, students should step outside of their formal education programs and take advantage of the any number of freely available online resources, such as MOOCs, online certificate programs in specialist areas of the field, and use online tutorials to explore aspects of the geospatial ecosystem not available within their chosen program. In particular, it is important that students investigate models of the geospatial sector, such as the DOLETA GTCM, and research the GIS&T Body of Knowledge. Students should also learn the sector they hope to become part of and understand it to prepare themselves appropriately for transitioning to work.

It is also important to consider the impact that these student- and program-focused recommendations will have on the nature of geospatial instruction in the future and to ask important questions around this. For example, how should university programs best be constructed to provide the best possible and most rounded intellectual learning environment for students? Or, should university programs and college programs in geospatial science be undifferentiated in content and purpose? Should, as is happening in several parts of Canada, there be agreements forged between universities and colleges in geospatial science where students can study in one environment for 3 to 4 years and then move to the other environment for 1–2 years of "finishing," to be immersed in deep technical use of geospatial science software and techniques?

While answers to these questions are interesting to ponder, for those currently studying geospatial science, as well as those about to enter courses over the next half decade, it is important to plan and prepare properly for the journey ahead.

Acknowledgments The authors are grateful to colleagues who shared their personal job search stories and advice. We also acknowledge the assistance of Ms. Katie Morehead of URISA, who provided access to the salary surveys, and Ms. Lisa Lin of Career Planning Service at McGill University for suggesting Canadian Government resources.

References

ACIL Tasman. (2008). *The value of spatial information: The impact of modern spatial information technologies on the Australian economy*. Canberra, Australia: Cooperative Research Centre for Spatial Information. https://www.crcsi.com.au/assets/Resources/7d60411d-0ab9-45be-8d48-ef8dab5abd4a.pdf. Accessed 8 November 2018.

ACIL Tasman. (2009). *Spatial information in the New Zealand economy; realizing productivity gains*. Melbourne, Australia: ACIL Tasman Pty Ltd. http://www.linz.govt.nz/about-linz/our-location-strategy/geospatial-projects/spatial-information-new-zealand-economy. Accessed 8 November 2018.

Association of American Geographers (AAG). (2018). *Guide to geography programs in the Americas 2017–2018*. Washington, DC: Author. http://www.aag.org/galleries/guide/20172018_Guide_to_Geography_Programs_in_the_Americas.pdf. Accessed 8 November 2018.

Association of Universities and Colleges of Canada (AUCC). (2011). *Trends in higher education: Volume 1 – Enrolment*. Ottawa, ON: AUCC. https://www.univcan.ca/wp-content/uploads/2015/11/trends-vol1-enrolment-june-2011.pdf. Accessed 8 November 2018.

Borwein, S. (2014a). *Bridging the divide, part I: What Canadian job ads said*. Toronto, ON: Higher Education Quality Council of Ontario. http://www.heqco.ca/SiteCollectionDocuments/Skills%20Part%202.pdf. Accessed 8 November 2018.

Borwein, S. (2014b). *The great skills divide: A review of the literature*. Toronto, ON: Higher Education Quality Council of Ontario. http://www.heqco.ca/SiteCollectionDocuments/Skills%20Part%201.pdf. Accessed 8 November 2018.

Canadian Council of Chief Executives (CCCE). (2014a). *Second survey report: Skills shortages in Canada.*. http://www.ceocouncil.ca/wp-content/uploads/2014/03/Second-survey-report-skills-shortages-in-Canada-13-March-2014.pdf. Accessed 8 November 2018.

Canadian Council of Chief Executives (CCCE). (2014b). *Preliminary survey report: The skill needs of major Canadian employers.*. http://www.ceocouncil.ca/wp-content/uploads/2014/01/Preliminary-report-on-skills-survey-Jan-20-2014-2.pdf. Accessed 8 November 2018.

Canadian University Survey Consortium. (2015). 2015 Graduating university student survey. http://www.cusc-ccreu.ca/CUSC_2015_Graduating_Master%20Report_English.pdf. Accessed 8 November 2018.

Canadian University Survey Consortium. (2016). 2016 First-year university student survey master report. http://cusc-ccreu.ca/?page_id=32&lang=en. Accessed 8 November 2018.

Coates, K., & Morrison, B. (2012). The uses and abuses of university: Canada's post-secondary system is failing its students and the economy, *The Walrus,* October issue. http://thewalrus.ca/the-uses-and-abuses-of-university/. Accessed 8 November 2018.

DiBiase, D., DeMers, M., Johnson, A., Kemp, K., Luck, A. T., Plewe, B., & Wentz, E. (2006). *Geographic information science & technology body of knowledge*. Washington, DC: Association of American Geographers.

DiBiase, D., Corbin, T., Fox, T., Francica, J., Green, K., Jackson, J., Jeffress, G., Jones, B., Jones, B., Mennis, J., Schuckman, K., Smith, C., & Van Sickle, J. (2010). The new geospatial technology competency model: Bringing workforce needs into focus. *URISA Journal, 22*(2), 55–72.

Dietsche, P. (2013). Career planning in Ontario – Grade 10 students: Counsellor perspectives. *The Canadian Journal of Career Development, 12*(1), 4–16.

Employment and Social Development Canada. (2016). Canadian occupational projection system (COPS), search for occupational projection summaries (2013–2022). http://www23.hrsdc.gc.ca/occupationsummarydetail.jsp?andtid=42. Accessed 8 November 2018.

Ferguson, J. (2014). What can I do with my degree? https://ceric.ca/2014/10/what-can-i-do-with-my-degree/. Accessed 9 November 2018.

Friar, J. H., & Eddleston, K. A. (2007). Making connections for success: A networking exercise. *Journal of Management Education, 31*(1), 104–127.

Forret, M. L., & Dougherty, T. W. (2001). Correlates of networking behavior for managerial and professional employees. *Group and Organization Management, 26*(3), 283–311.

Gaudet, C. H., Annulis, H. M., & Carr, J. C. (2003). Building the geospatial workforce. *URISA Journal, 15*(1), 21–30.

Gedye, H., Fender, E., & Chalkley, B. (2004). Students' undergraduate expectations and post-graduation experiences of the value of a degree. *Journal of Geography in Higher Education, 28*(3), 381–396.

GeoCommunity. (2014). GIS career salary survey. http://careers.geocomm.com/salarysurvey/. Accessed 3 February 2015.

Gouvernement du Québec. (2013a). *Reliance survey of university graduates/2013*. http://www.education.gouv.qc.ca/fileadmin/administration/librairies/documents/Ministere/acces_info/Statistiques/Enquetes_Relance/Universite_Bac_Maitrise/Relance_universite_Bac-Mai_Brochure_2013_VA.pdf. Accessed on 10 November 2018.

Gouvernement du Québec. (2013b). *Relance survey of college graduates technical training/2013*. http://collections.banq.qc.ca/ark:/52327/bs2326858. Accessed 8 November 2018.

Helyer, R., & Lee, D. (2014). The role of work experience in the future employability of higher education graduates. *Higher Education Quarterly, 68*(3), 348–372.

HECSU/AGCAS. (2010). What do graduates do?. http://www.hecsu.ac.uk/assets/assets/documents/lmi/WDGD_Nov_2010.pdf. Accessed 8 November 2018.

Job Futures Quebec. (2015). *Mapping and related technologists and technicians*. http://noc.esdc.gc.ca/English/NOC/ProfileQuickSearch.aspx?val=2&val1=2255&ver=06. Accessed 10 November 2018.

Johnson, J. (2010). What GIS technicians do: A synthesis of DACUM job analyses. *URISA Journal, 22*(2), 31–40.

Koss-Feder, L. (1999, March 22). It's still who you know … in the boom economy, job hunting is a way of life. Here's how to do it. *Time, 153*(11), 114F.

Linnitt, C. (2013). *Harper's attack on science: No science, no evidence, no truth, no democracy.* https://academicmatters.ca/harpers-attack-on-science-no-science-no-evidence-no-truth-no-democracy/. Accessed on 07 December 2018.

McKendry, J. E., Adams, J. K., & Solem, M. (2014). Employer perspectives on the value of geographic expertise, skills, and technologies. *Research in Geographic Education, 16*(1), 23–39.

Monk, J. J., Foote, K. E., & Schlemper, M. B. (2012). Graduate education in the U.S. Geography: Students' career aspirations and faculty perspectives. *Annals of the Association of American Geographers, 102*(6), 1432–1449.

Natural Resources Canada. (2015). Canadian geomatics environmental scan and value study. http://giscourses.net/wp-content/uploads/2015/05/Canadian-Geomatics-Environmental-Scan-and-Value-Study.pdf?189db0. Accessed 10 November 2018.

OECD. (2013). *Education at a glance 2013: OECD indicators*. Paris: OECD Publishing. https://doi.org/10.1787/eag-2013-en. Accessed 8 November 2018.

Refling, E., & Borwein, S. (2014). *Bridging the divide, part II: What Canadian job ads produced.* Toronto, ON: Higher Education Quality Council of Ontario. http://www.heqco.ca/SiteCollectionDocuments/Skills%20Part%203.pdf. Accessed 8 November 2018.

Sattler, P., & Peters, J. (2012). *Work-integrated learning and postsecondary graduates: The perspective of Ontario employers*. Toronto, ON: Higher Education Quality Council of Ontario. http://www.heqco.ca/SiteCollectionDocuments/WIL%20Employer%20Survey%20ENG.pdf. Accessed 8 November 2018.

Schlemper, M. B., Adams, J. K., & Solem, M. (2014). Geographers in business, government, and nonprofit organizations: Skills, challenges, and professional identities. *The Professional Geographer, 66*(3), 480–492.

Smith, P. & Lam, M. (2013). 2013 *Campus recruitment report recruiters summary*. Canadian Association of Career Educators and Employers. http://www.cacee.com/_Library/docs/2013_CACEE_Campus_Recruitment_Report_-_Recruiter_Summary_Final.pdf. Accessed 8 November 2018.

Solem, M., Foote, K., & Monk, J. (Eds.). (2013). *Practicing geography: Careers for enhancing society and the environment*. Boston, MA: Pearson.

Stigler, G. (1961). The economics of information. *The Journal of Political Economy, 69*(3), 213–225.

Tamburri, R. (2014). Co-op programs are popular and growing at Canadian Universities. *University Affairs*.. https://www.universityaffairs.ca/news/news-article/co-op-programs-are-popular-and-growing-at-canadian-universities/. Accessed 8 November 2018.

Turner, C. (2013). *The war on science: Muzzled scientists and willful blindness in Stephen Harper's Canada*. Toronto, ON: Greystone Books.

Urban and Regional Information Systems Association (URISA). (2007). *URISA salary survey for IT/GIS professionals*. Des Plains, IL: URISA.

Urban and Regional Information Systems Association (URISA). (2011). *URISA salary survey for IT/GIS professionals*. Des Plains, IL: URISA.

Urban and Regional Information Systems Association (URISA). (2017). *2017 URISA GIS salary survey*. Des Plains, IL: URISA.

Wolff, H.-G., & Moser, K. (2009). Effects of networking on career success: A longitudinal study. *Journal of Applied Psychology, 94*(1), 196–206.

Zhou, Y., Smith, B. W., & Spinelli, J. G. (1999). Impacts of increased student career orientation on American college geography programmes. *Journal of Geography in Higher Education, 23*(2), 157–165.

Epilogue: A Future of Convergence and Crises

James Boxall

We look at the present through a rear-view mirror.
We march backwards into the future.

Marshall McLuhan, The Medium is the Massage, 1967.

Abstract Teaching and learning GIScience, or anything for that matter, is influenced by the context of technology. The context is not the technology itself, as a tangible entity; rather it is the overall qualities, actions, people and outcomes which take place during current developments and that have a lineage to past shifts in technology. This chapter takes the last 60 years and breaks them into three periods of massive change: 1957–1995, 1995–2015 and 2015–2030. While the chapter outlines positive results from technological change that impacted GIScience, it does use a cautionary tone when considering the present lack of a shared goal. In some ways, the context today and moving forward is unique in that it may not be a context at all.

Keywords Education · Technology · Change · Space race · Prediction · Trends · Culture

Introduction

By 2025 we will have collected 163 zettabytes of data generated by human to human and human to machine interaction. What's in a zettabyte anyway, and why does it matter to learning and research in GIScience? A zettabyte is 10^{21} bytes. That's a trillion gigabytes. If one were to convert every element of human communication to

J. Boxall (✉)
Dalhousie University, GISciences Centre, Killam Library, Halifax, NS, Canada
e-mail: james.boxall@dal.ca

S. Balram, J. Boxall (eds.), *GIScience Teaching and Learning Perspectives*, Advances in Geographic Information Science,
https://doi.org/10.1007/978-3-030-06058-9_10

binary format, then by 2025 there will be enough storage capacity to house every written and spoken word, for all time (Cave 2017).

Speaking of trillions, Apple Inc. is worth a trillion dollars. To explain it in terms we can all appreciate (and envy), take fresh, new $100 bills in neat $10,000 bundles, place them in tidy stacks 7 feet high, and then cover a regulation football field with those 7 foot stacks. That's a trillion. This metaphor sheds light on a fundamental issue we face in learning and research throughout academia. Zettabytes affect more than our spatial work and community; they may be making the future impossible.

In looking to the future of the intersecting disciplines discussed in the previous sections of this book, the first thing one notices is that the future is more unpredictable based upon where we have come from and what we see before us today. We hope we can come close in our estimates of specific technologies or methods. However, we are likely to miss the mark by a fair amount. We tend to look at the physical, tangible product or service rather than looking at the context which will shape every change.

At its core, GIScience is about data, technology and analysis as it relates to geographically referenced information – a greater emphasis on those tangibles above. While elements of GIScience can be (and are best suited to) nondigital environments, it is impossible to think of GIScience being done – on the whole – without accepting that it is reliant upon digital data and high performance computing or big data and supercomputers supported by a cloud infrastructure. So if we accept that GIScience is a digital enterprise, then anything which affects its present use and future iterations should be reflective of all computing, data and communication developments, past, present and future. The zettabyte is becoming the metaphor that will shape the context for our futures.

The previous chapters set forth some vital considerations for those practicing, researching and teaching GIScience, and the authors should be congratulated for highlighting the foundations of our discipline as it was and is, along with aspirations for future directions in research and practice. As colleagues, we see each other's perspectives and appreciate the desire each has for the advancement of the discipline, with the resulting benefits to our students both directly within classrooms and indirectly through support within the private sector and government. After all, if it does not impact our students (and through their future contributions, society), then there really is no rationale to support our efforts.

One thing I consider to be at the core of GIScience research and learning is technology. It is the primary intermediary for what we do. It goes far beyond mere tools. It extends into decision support and solutions requiring spatial analysis, and which traverse any knowledge domain, and at every scale imaginable. We accept this within formal teaching settings, in research and for application development that is closely aligned with computing, human computer interaction and visualization. Even the creation of new theories, methods, curriculum and pedagogy is dependent upon or designed for what is possible in a digital environment and mediated by software and data access.

The following comments are problematic only in so much as they are blunt. They begin situated within a time when developments in computing and geospatial data followed a pace where the mere acknowledgement of such a thing as a zettabyte would have been inconceivable. Changes in the digital landscape should, therefore, help us determine the most likely path along which we can adapt what it is that we do. These concerns are based upon trends (and actions) that are moving as quickly as the processor embedded in my laptop where these words are placed. This points to another great issue we face: change is not linear, and shifts in the *rate* of change will have feedbacks full of unintended consequences we would not have wished for. We are almost capable at planning for linear change; we are far less successful for exponential and/or multidirectional adaptation.

Acceptance of technology through blind faith is dangerous. We should recall that in the 1950s, cigarette advertisements had the endorsement of medical doctors and were sponsors for all manner of TV shows and sports events; thalidomide was a cure; and asbestos was a miracle to be used in any home or office. Imagine the person who, late in 2006, based upon the best of knowledge and research, invested all their money into the flip phone (or even BlackBerry). Who amongst us could have known the iPhone would be available in 2007? This was not the first time, nor would it be the last, when change was unexpected, unpredictable and came at a speed no one could imagine – with fallout measured in substantial crises for businesses and massive shifts in society. Recall that the Pony Express came into being in 1860. It collapsed 18 months later as the Morse telegraph joined the North American continent by wire, and communicating from coast to coast suddenly became infinitely faster, cheaper and safer. A rule in life is that it can change in an instant; the context that surrounds our lives makes those instants possible.

The value society puts on education is determined mostly by the demands and desires of the 'times we live in'. Recognizing the importance of *context* is perhaps one of the few things all educators can agree upon, along with the reality that society seldom recognizes the real value of learning. It does, therefore, make ample sense to look closely at what will likely influence the direction of GIScience education based upon context and trends. Education and research exists on a spectrum that shifts quickly due to the whims of funders and the ever-changing public-political expectations. Even the best ideas created within the learning enterprise will gain little traction and may not even see the light of day if they do not fit the current desires of the non-experts. Education is one of the most contested and essentially political systems we know. These socio-political drivers of context are critical to recognize. Moreover, it is equally important to keep in mind that pressure now comes from the convergence of the external social forces, our own desires and the increasing speed and depth by which our lives are becoming more digitally integrated.

Context

This is not something that can be isolated as 'what is happening today'. It is more fluid and nostalgic. It takes into account how one arrived at the 'now' and the likely future our actions today will influence as the 'later'. If you want to build a flying car in the future, the context of today would tell one if it would be of use. You would also have to consider a linage that can be traced back to the Wright brothers or even to da Vinci and his drawings of human-worn wings and helicopters. Then again maybe it goes back further to the myth of Icarus and our resulting dreams over millennia to fly. The 'now' and 'later' are informed by the 'before'.

Context is always how we connect the past to the future; by definition that is the essence of context. Our differing perceptions refine the various perspectives we have about those connections; how I see the development of the Internet of things (IoT) and its future will be slightly different from yours. How we see the impact and benefit of the IoT will be equally unique, but likely very much more in conflict. Those conflicts are likely to become very evident over the following sections of this 'last word'.

To look forward with regard to the progress and timeline of technology is muck like a fool's errand. Two years feels like a lifetime; 10 years is akin to a century. To know what is to unfold – and the impacts – is nearly impossible. Therein is the quandary, even the trap. My views have the qualities of Schrodinger's cat; they both exist, and do not, at the same time. If it is impossible to know what technologies will arise, likely shocking us in unexpected ways, then planning (funding, shifting curriculum, etc.) for education should be viewed with some suspicion for, as we will see, knowing how things are used seldom follows the dreams or designs of creators and users.

Regardless of the above, or what is to follow, this epilogue is not presented as 'antitechnology' nor anti-anything per se as related to what we do in our research and teaching. This is a deep look and reflection on where we 'may' be going. As all the authors in this book have at the core of their thoughts the nature and delivery of education in one way or another, then it would be ridiculous to not apply the same concerns to this final chapter. From a teaching and learning standpoint, this final text was one that demanded more reflection than research. It is situated in the experience of seeing a major, unprecedented, hyper-speed shift in student learning capabilities and outcomes dominated by the digital technology in their hands.

This epilogue positions us, in Aristotelian terms, at the centre point – the convergence – where *Techné* (craft), *Episteme* (knowledge) and *Phronesis* (application/solution) are overlapping entirely. The convergence of these three aspects of learning is being driven by – and for – technology rather than any socio-environmental good. This convergence is itself becoming the problem. As Gazzaley (2018) suggested, we are facing what may be a dangerous 'cognitive crisis'. This moment is one when we need to step back and recall where we came from, where we are and the *qualities* of the path(s) that lay before us. As McLuhan noted, we have to see the past at the same time as we move forward (McLuhan 1967).

This final reflection uses three sections in our technological history as anchors for analysis and prediction. It moves from past (1957–1995) to recent (1995–2015) and then to present and future (2015–2025). To each timeline, there is a theme added as a tool to focus on outcomes of that time, all of which reflect a personal bias. We begin at one of the most pivotal moments in the history of science and technology. It was a time that highlighted how a human crisis, made possible by radical – even shocking – shifts in technology, could result in unexpected outcomes. One such result was that it became a time which seemed to ask 'why' rather than 'why not'. We will see that those simple questions will change as time goes on.

Theme One: Community and Control (1957 to 1995)

On the surface, one might rightly question a theme for this period of time which is a paradox when we juxtapose the ideas of 'community' and 'control'. It was certainly not the development of a global community, although scientists tended to try and overcome barriers to share work, simply for the sake of science. It was about creating communities of practice, within the communities of political commonality. It was a period of funding increases (outside of WW2) that would last longer than any global war. It began as a time when success was measured by a suburb, ironically a community controlled by the car, the road and commuting. It ended during a decade that seemed to be everything for everyone. That would soon end.

What sparked this time? Simply put, it was a little sphere, no more than 58 cm in diameter. On October 4, 1957, Sputnik1 soared into the skies and began to orbit the Earth. The tiny 'beep, beep, beep' sound it made created fear in the US government (and others) and a degree of curious panic across society as a whole. If they can put a transmitter into space, why not a nuclear bomb? The space race was on. What most people ignore is that 1 month later, a dog named Laika went into space aboard Sputnik2. While Laika died during the mission, she did survive the launch and first stages of flight, thereby proving that a living organism would be able to survive, if no mechanical errors happened.

The start of the human space race was on, and the move towards more STEM education and research became paramount. Massive investments were made; anything to help the cause was deemed worthy of a try; failures became numerous and the public and funders became annoyed when rockets exploded on the launchpad; the race was being lost for the 'west' – for a time.

From the perspective of GIScience and allied disciplines, this period gave us the basis for four developments that are of greatest impact and lasting contributions: space-based technologies (orbital ability), GIS and the quantitative revolution across geographical sciences, advancements in computing and the development of the Internet.

Moving forward from 1957 to September 12, 1962, when US President John F. Kennedy gave a speech at Rice University in Houston at the site of the NASA command centre, we come to see how deep the desire to win had become.

> We choose to go to the Moon! We choose to go to the Moon in this decade and do the other things, not because they are easy, but because they are hard; because that goal will serve to organize and measure the best of our energies and skills, because that challenge is one that we are willing to accept, one we are unwilling to postpone, and one we intend to win, and the others…

Kennedy's speech was important because it raised the profile of NASA. It was even more vital because it captivated a nation; it gave everyone a sense that they were special and can achieve great things. Most importantly, it stated that they had to 'win'; it highlighted the reality that the Cold War was in space as much as it was anywhere else. The impact of this was that just a little more than a decade from Sputnik, NASA would fulfill Kennedy's challenge. Personally, the call to do the hard things to rise to a challenge beyond oneself resonates deeply, as it has since that time.

What was needed to create this technological leap? Simply put, over 400,000 people were employed in the space program. Thousands of companies were formed, and children everywhere were keen on science in schools, as they 'wanted to become an astronaut!' From the perspective of both space adversaries of the Cold War, success required the largest focused communities akin to previous global wars. It also, in similar fashion, required the greatest degree of control as 'failure was not an option'.

One should also recall that Gene Kranz, Mission Control Flight Director, remarked that when Apollo 11 landed on the moon on July 20, 1969, the average age in the control room was 26 (PBS 2016). Oddly enough, it may well be true that this could be the average age in many software developer 'control rooms' today. In addition, if the average age were 26, then these people would have been 14 or so when Sputnik occurred. These would be the first beneficiaries of the increased focus on the sciences and technology in high school and university. It seems that purposeful change works, while lacking a purpose, even control, does not lead to work.

As an aside, all three astronauts who went to the moon on Apollo 11 were born in the same year – 1930. All were part of the silent generation – those who did not question authority but merely went and did the work as expected – another 'control' aspect of the theme. The next generation would not be so silent. That reaction to control would have a huge impact on the foundation of computing and the Internet. These are the things to drive the next phase. The 'norms' of the decade after World War Two were about to become challenged and overturned by the baby boomers – the largest cohort, with the most money and education ever.

Up to 1969, at least, science had won. This was a time when the United States was winning the race. The spillover into other countries was beginning to come to the fore quickly (Germany, UK, Japan, France, USSR and even Canada) as satellites and advanced communication systems were so vital to the development nations to lead in GIScience, RADARSAT, Canadarm, GEOS and every sensor platform, the Space Shuttle and the ISS. This was the heyday, and it was about to get bigger, much bigger, and very quickly. They can all trace their birth to 1957.

Two gigantic shifts occurred that would have lasting effects on our sector. They are perhaps the only transformative things in the truest sense of the word. The creation of GIS and the development of the Internet (DARPAnet) both came alive at the time of the moon landing in 1969. This was the year when the seeds were planted for what would become all that we do in the technical aspects of our research and learning in GIScience. From a geospatial point of view, 1969 was an exceptional year – it was turning the corner around the path we would follow for the next

50 years. Of course there are numerous other developments being glossed over here, and so much of the history of this time (and the ones to follow) is avoided. A limit of space forces one to pass by even some of the most amazing events.

From this moment grew wider Internet access as universities began to create more services and protocols were developed and made available – that is, if you were lucky enough to be connected to the backbone of the network. Campuses and research labs everywhere would eventually connect, and this meant a larger, interconnected community of researchers that could begin to work together across a distance and faster. This was building a science and technical community in a way that helped maintain all our communities (Kohn 1970).

While these developments brought us to 1969, the years from 1970 to 1995 were equally special in the development of the spatial sciences. We were entering the commercial success period, which also shaped GIScience. People wanted science, they wanted to trust scientists, and they did. This was a period where consumers trusted what they ate, were prescribed, smoked and had around them in the air and water. But the closer we get to 1995 and beyond, those qualities and attributes decline, slowly at first.

In GIScience (and related areas), there was a fortunate mix in this period. Both the root disciplines – geography and computing – were recognized and appreciated, and the private sector was supportive and growing rapidly. Universities were active and growing rapidly to support the needs internal to the campuses as well as wider societal needs. The development of better sensors for imagery, better surveying technologies, GPS, spatial analysis models and statistics along with the improvement in computing also supported the development of the often-undervalued areas such as spatial data infrastructures, standards, policies for sharing data, models and the like. Communities were growing and working together. Controls were in place to make certain it could come together while still protecting individual goals and business needs.

Moreover, in a little more than a decade, the external pressures on education and research had changed radically, thus causing tectonic shifts in the academic world, specifically with program development, research and funding. In geography, we saw the 'quantitative revolution', which meant an increased focus on spatial theory and geostatistics. This was largely due to a desire to be a part of STEM and 'of enough value to gain research dollars' as was in keeping with the tradition of the space race. Funding went to those who did math, models and machines and not description of landscapes or pretty maps. That is an overgeneralization, but it is meant to highlight how the larger shifts or external pressures can cause changes that spread across other domains or disciplines. The questions today are the following: are we seeing this once more, and will it continue into the future (like a decade such as the mood shot)? What is the Sputnik of today?

From 1970 to 1977, we saw the beginnings of the personal computer revolution. From 1980 to 1983, the home computer became more accessible, thanks to the creation of PC clones to keep prices down. In 1983 a small company called Apple released Lisa, its home computer. To correct mistakes and design flaws in Lisa (and Lisa2 and the MacII), the Macintosh was released in 1984. It was genius that Apple used the *1984* of George Orwell in the advertising campaign for the Mac which became one of the most recognizable and popular ads ever – 'So 1984 won't be like 1984'. Computers became smaller and faster and now with laptops are becoming more popular and possible with better chips. Cell phones were on the horizon, even if they were large and limited in functionality (with unreliable range and many dead zones).

Additionally, 1983 was the year when *Time Magazine* made the computer the 'Machine of the Year', replacing what was normally the annual 'Person of the Year' special issue. The age of the computer had arrived. Now came the moment when computers had to generate applications in order to grow in status and usefulness, and connectivity with others was a hope. This was also the time of a science fiction author, William Gibson, who coined the term *Cyberspace*, a word to help define what would be a shared lexicon grasped by the public – a public so wide it would span the globe and take us into another stage of dominance by the machine.

Following on from the quantitative revolution, our communities were positioned to see the next increase in speed and the acceptance of a society newly joined by the computer and the Internet, with emerging disciplines ready to capitalize on these new tools. Applications became essential to develop as processor chips became more powerful. GIS was growing quickly in organizations that could have large workstations with massive power and a plethora of people to do different functions. GIS needed power, and the chips were not there yet. Things were about to change. At the same time, GPS was growing and becoming more accurate. The literature was suggesting a near future which would see things like GPS used in everything. How often do we underestimate how technology will be used?

At the same time, computer science programs expanded as did those in the engineering sciences; rocket science was still something one could bank on. New geography programs started at universities, while research, publications and grants increased. Associations, publications and new specialist communities were forming. The computer and GIS (sometimes computer cartography) were key areas that would define the status of a university department and the future of how GIScience would be incorporated into research and learning. Those at the leading edge would be met with growth.

Advances in computing meant GIS could improve in ways that would support technical spinoffs (innovation), new companies and more research. The vitality of the computer is the vitality of GIScience. Even for research and learning, there was one thing no one could have predicted that would change the landscape forever. There was hope. The hope was based upon communities growing, developing new tools and ideas and a feeling of being at the start of a new revolution. That outcome would have been beyond anyone's ability to forecast or predict back in 1957. Desktops meant that people could own and work with computers, software and data from home. The CD made working with the computer and sharing data or software easier and easier. Excitement followed hope, even if the power of the technology was insignificant compared to today. Speed, power, capacity and connectivity are relative. Technology is situated in a time and a nostalgic context of the users and creators.

From 1957 to 1995, in very broad terms, the computer revolution took place. It was essential for geospatial sciences that relied upon tech and data, be it from more accurate surveys or from space-based sensors. This was also the science revolution where we still needed to win the race because it was also the Cold War. It was the time when we became a community seeking answers, not merely accepting and marching forward as our parents did. We depended upon experts who had access to technology and who proved their value through grand experiments like the moon landing and consumer products we could easily obtain. Money was available to 'buy' the good life, consumers were the focus of everything, science was king, and the computer was starting to be seen as the next 'must have'. This was our world, for now.

There was one thing, however, that began during this phase and was still under some control or within reach of our understanding. We were about to face a data/information flood that was not planned for nor fully understood in its implications. We had too many new sources of data, too much being collected, and it was becoming hard to process and find a use or even the answers to what was originally the purpose to collect it all. It was and is an irony that spatial analysis can always use more data. However, the speed of this information explosion is not just data but ideas and other communication forms we rely upon.

As the founder of the National Research Foundation, Vannevar Bush, stated in 1945 – clearly prescient and certainly worth recalling:

> The difficulty seems to be, not so much that we publish unduly in view of the extent and variety of present-day interests, but rather that publication has been extended far beyond our present ability to make real use of the record.

Little did we know that the next leaps would be more than Vannevar could have ever dreamt of. He was thinking of a time when books and journals were growing in number, but libraries were not expanding at exponential rates. The volume of data growth and converting just print to digital would be less than 10 terabytes with a growth of 2 MB for each book. At that volume, I could easily store the Library of Congress in my lab – three times over. If it was impossible to engage that amount of information in 1945, how do we survive now? Did we lose or gain something? What if 90% of all data created or used is junk, never to be seen, taking up space? We created the computer, and it satisfied our appetite with bytes and bytes.

As we will see, the next phase from 1995 to 2015 would bring forth change that would surpass that of the space race. This may become viewed as the period that makes all technological change and revolutions prior pale in comparison.

Theme Two: Creation and Consumerism (1995–2015)

It is clear that 1957 was a year when a singular event created important reactions that sparked a surge in the development of digital technology, theory and practice. While the starting point for a second phase of the context surrounding GIScience learning and research could be almost any year, 1995 has special significance for our domain. Had it not been 1995, I may well have chosen 1989 due to the Berlin Wall coming down – in essence the end of the Sputnik era.

This 'year' (using a bit of poetic/temporal licence) saw the time when Tim Berners-Lee reached the public online and sparked the creation and distribution of browsers of such as Mosaic (1993), Netscape (1994) and Internet Explorer (1995). Additionally, AltaVista and Yahoo began during 1995. It was a year that marked the start of something, even if people didn't know exactly what that was or where it would take us.

These became our entry points to navigate the Internet (the 'web', the 'information superhighway'). We needed it to be easy to get at more content from our homes via Internet connections now supported by better modems. We could then view anything on our superfast computers and even on our laps in the bedrooms. With the immense increase in modem speeds over these early years of 'dial-up Internet', we saw the potential to deliver imagery more quickly, and the concept of online video chats and streaming of movies became things in development rather than in dreams. In 1998, the movie *You've got Mail* brought to life the public acceptance of emotion, even love, being available online. Our connections to each other became virtual as much as real; online dating sites grew into big businesses and very quickly. With the power of the new Pentium chip and DVD came a business model for Netflix (1997).

Because of the technologies associated with a web-based Internet, the planning, development and implementation of online education began to take hold. Everything that had seemed fanciful was now seen as possible. We saw in our domain the ability to expand the Geographer's Craft begun in 1991 by Colorado University professor Dr. Kenneth Foote. It was a brilliant idea and super effort, but it would be overtaken by technologies to come later, even if those did not have the same sense of community (perhaps a legacy of the previous timeline). Predictable? Many make a living at being futurists. They only need to be right once.

For the geospatial community, 1995 was a busy year and the start of two decades of tremendous and unprecedented growth. We saw ArcView1 as the first Esri release of software accessible by anyone with a personal computer instead of a workstation. Clark labs released IDRISI which allowed educational institutions to do more in their labs and to extend into raster-based learning. There was also the Alexandria Digital Geospatial Library (ADL) project out of the University of California, Santa Barbara. Alexandria was the effort to explore the dream to develop the most robust platform for storing, finding and accessing geospatial data (ADL 1995). We can thank ADL's technical and process outcomes for much of what we use today that seems both obvious and invisible in our processes.

This was a time when we were solidifying a new culture and desire to get data and software out of the lab and into the hands of more people, more students and researchers and even those simply wanting to explore their world. As more and more of this began to take hold, we saw a new concept arise – VGI or volunteered geographic information (Goodchild 2007). All of this meant a more concerted effort to expand, enhance and simplify the spatial data infrastructures around the world. This became so critical a foundation to our location-based industries that the UN

Global Geospatial Information Management (UNGGIM) was formed. Moreover, one cannot ignore the role of standard organizations like the Open Geospatial Consortium (OGC) who navigated the issues of sharing data to make certain ISO standards appropriate. Suffice it to say, these were busy days, creating what we use today.

All of this led to the other aspect of this theme of creation and consumerism. The consumption of services like Google Earth and Google Maps, Bing, Waze and every type of location-based service increased exponentially with the incorporation of GPS and navigation tools in cars and soon to be in phones. These went from the toys of the rich in expensive vehicles to the common accessory one is shocked when it is not present. This was the point at which the technology of position passed the position of people to understand their technology.

Competing search engines, from both new companies or old companies, seek a place in a new type of market, all updating constantly to have systems keep up with tech, and vice versa, so as to ensure the consumer was satisfied and that the market grew. Competing and contested landscapes within the discipline(s) were the point of departure from GIS acting alone or geography acting as an umbrella. The very terms we used to call what we did became a point of confusion and a bone of contention. This was the time when GIScience was born in an attempt to overcome the limitations of previous definitions and understanding of what we do and how it affects society (Goodchild 2009). It was a time when journals changed their names, such as GISystems (1987–1996) to GIScience (1997–). These were the years when the search to create the Digital Earth described by Al Gore (or the speech that never happened) became reality with the International Society for Digital Earth (Gore 1998).

These were (and are) the years when anything seemed possible, when we saw growth in the sector even when downturns took place in the economy. Creation was at breakneck speed. The consumption of these products and services was at a level where one could not see an end over the horizon. It also, sadly, was an era of constant updates to software and hardware, at times requiring new hardware simply to start the software, or hardware that cannot update using the OS of the company predicting the next 1984ish hardware (we have all been there, and need not mention the company). We created computing in the era before 1995. We became experts at finding every possible use for those developments of the past. We became consumers waiting with baited breath for the latest product or service that was to be released.

And like Sputnik, there were still geopolitical realities front and centre also encouraging expansion. We began to see new forms of international conflict that were not based on Cold War alliances; rather they were breakdowns and breakups of old conflicts. We first came in contact with the term ethnic cleansing. It must be noted that the many horrific events of this period should not be so cavalierly passed over with the above summary of this global period. It begs the question I've asked myself, and for which I have never answered: As the East-West geopolitical fabric began to unravel at the same time the Internet took off, and distracting technologies were served to the public, did we experience a convergence, where the complexity of the world was met by a simplicity of online life?

There can be little doubt that the Sputnik era spurred on a tremendous amount of military investment and development. This period of change in all things digital was both impacted by, and an influencer of, greater incorporation of geospatial. This period saw crisis after crisis, matched by new technology (who can forget the phrase smart bomb) and now drone pilots wearing flight suits and launching aircraft from 15,000 miles away from their targets. It is not only engineering and aeronautics but also GIScience that supports the new forms of warfare. We support the application of surveillance and locational technologies for tracking people in the real and virtual worlds. We are part of the 'military industrial complex' President Eisenhower warned of. Of course, conflicts arising from 9/11 as well as other crises that have included a new phrase – ethnic cleansing – cannot be ignored. This period of time was not peaceful by any means.

Then 2004–2007 happened. Facebook and the iPhone. Social media in a mobile, web environment changed everything. If 1957 was the creation of STEM, 1995–2015 would create the rationale for a totally different, larger and more consumer-focused STEM.

If this is a period of creation, then it can best be summarized by the term "app" to describe the technologies added to our phones everyday or those we share across the Internet built over the previous decade. No one saw this coming, because no one saw the smartphone on the horizon, even if Captain Kirk's communicator gave us a subtle hint of where to boldly go. But these apps summarize the miniaturization of all location-based technologies. It is almost funny – we have made the world as small as the phone. The downside can be expressed as simply as the mobile apps we now use to generate mobile orders that allow us to cut in line and receive a service simply because we ordered via a location-based technology. A new digital divide has formed based upon who has what app.

This divide is about to be made more noticeable. The growth of, or attention towards, the Internet of things, machine learning, artificial intelligence, individualized apps, new social media platforms more enticing than previous ones, open data built on VGI, STEM education mandatory coding courses in schools, data ownership and location privacy, copyright, streaming, political influence of platforms on social media, drones, big data and analytics all point to the changes in the context of our sector and what we do. The pace of change is so fast it has become cliché.

One could go on and on. The issue here is not what to include but what to exclude. It is not a matter of how much to say, but when it is right to oppose, to say no more, or as Billy Joel wisely said:

> 'You have not put out a pop album since 1993….Why not give us more?'
> asked Stephen Colbert.
> 'I thought I'd had my say…and, well, ok, shut up now' (Colbert 2017).

The smartphone, the available data, the mobile apps and the new software have all made it possible to consider any possible outcome. It has become so critical to life that we see the need for mandatory coding in schools, so students will be ready to add to this landscape in the future. It raises an important question for those of us in GIScience. Is STEM designed to build new things to help us, or is it about creating things so we consume more? A reminder of this are the aims and goals of statements or reports such as Amplifying Human Potential: Education and Skills for the Fourth Industrial Revolution (InfoSys 2016). 'If we don't prepare students with the skills they need for today's job market, who will (Cassell 2018)?' This is the view we are facing. As much as we in the tech sector can be supportive, we have an obligation to look at these critically, thereby providing direction which tones down the wonderment and provides some realism. If not, we are just as culpable as anyone else for what we will face in the future. Suggesting there is a saviour on the horizon or a panacea of a solution is troubling, and it may be a solution looking for a problem.

Some time back during the creation of GIScience, David Mark suggested "GIScience also examines the impacts of GIS on individuals and society, and the influences of society on GIS" (Duckman et al. 2004). This, from my perspective, is hitting the point about context. It also highlights the need to take a careful approach and review all we do which the remainder of society see as wonderment without question. I would also suggest we have not spent enough time focusing on the second part of Mark's statement – how society (context) affects what we do.

If the growth in GIScience over the last two decades has been spurred along by the growth in the creation of new technologies – at an amazing rate – we must consider the reason. It has taken place along with an equally fast consumption of such technologies by an ever-growing technologically demanding and 'savvy consummerati'. If this period was such a time of creation and consumption, which followed a time of discovery and vision, then the next era must be equally unique.

Theme Three: Confusion and Conflict (2015 to 2030)

US patent number 9,280,157 B2, March 8, 2016. That number does not sound very exciting, nor is it likely to spark a memory in any of us. That patent was for a *System and Method for Transporting Personnel within an Active Workplace* (Amazon Tech Inc 2016). This patent was filed by Amazon as a means to have workers protected from the robots around them. It was a cage for people. Amazon later stated it was a bad idea, and it was to be replaced by vests worn by workers which caused the robots to stop moving if they were near each other. What a great relief (Dore 2018).

The World Economic Forum's (WEF 2018) *Future of Jobs Report 2018* and the McKinsey discussion paper on *Ai impact on the World Economy* (McKinsey 2018) highlight the changing nature of what we do and how we do it and how such change will take place over the next several decades. As the WEF puts it:

> The Fourth Industrial Revolution is interacting with other socio-economic and demographic factors to create a perfect storm of business model change in all industries, resulting in major disruptions to labour markets. New categories of jobs will emerge, partly or wholly displacing others.

How could any discipline or line of business (be that GIScience or GIS) cope with such a 'perfect storm'? Because of what we do – and how vital location is to life – we will do much better than most other sectors. No matter what ideas are expressed about this period (especially mine), from now into the near future, the one thing we can say for certain is that geospatial data, GIScience and geography (all taken broadly) will be the most common and essential parts of all technical, social and natural projects (Unwin 2011). It may be Ai and automated systems, or health, or climate, will be core areas to focus on for success. Having said that, the broader context within which our field will thrive will be problematic, having the potential to derail the best of opportunities. The context in the future may be one where few will be able to adapt to the needs of society, much less the development of new solutions.

Several concerns arise from this massive shift as described by both the WEF and McKinsey. First of all, the types of jobs that will be displaced and the types to emerge as beneficiaries are almost all in the technology/data and business/tech-service sectors. According to the WEF, 133 million new jobs will emerge by 2022, while 75 million will disappear. We therefore will see a net growth in jobs of 58 million over the next 3–4 years. This figure is, however, based upon an analysis of 20 countries (with a summary review of all regions). This raises the question as to where the jobs will be lost. The ability of economists and management consultants to predict and plan for labour change, reskilling and technology is not at a level to support the ideas of what will happen and the benefits/losses to come. Moreover, new certification and credentials will be required for those to be reskilled – some in areas that do not exist as of yet. Seldom do such mass changes in employment and labour structures meet with opportunities to find new or equivalent employment; too few win, and too many lose.

What will those new jobs be? What will the business lines be which are most dependent upon the contributions of GIScience? What focus do we have which will allow us to carve out a niche where GIScience supports other fields but also allows a niche to grow where GIScience can contribute on its own? Again, to try to plan for what will result from human systems changing over time is an effort best left to psychics. The most we can hope to do is understand the changing context today. Predicting further into the future and across or between systems – human, technical and natural – can only become reality if one is planning, designing and building all the right pieces just for that future. That is not luck; it is predetermination.

Based upon the current state of affairs, this next phase from 2015 to 2030 may not have an outcome – a new context – which will be supportive of the future. As the graph adapted by Freidman shows, the human system has fallen behind the technical, and we cannot adapt enough to keep up with the exponential pace of change (Friedman 2016). This one fact alone has massive implications for the future context which will be putting pressure on our discipline and educational efforts.

We have become tribal like never before, helped along by social media (Chua 2018). This divides us. Oddly, it also cuts down on confusion because we are 'hanging out' with those who are like-minded. The downside is that divisions are the basis for more and deeper conflicts when we encounter those from other groups. This is sad and does not bode well at a time when we need communities to work together. We are living in a time when the path we lay down before us, from now into the future, cannot bear to have such conflicts arise; this is an era when tribalism needs to be pushed aside, not reached out for.

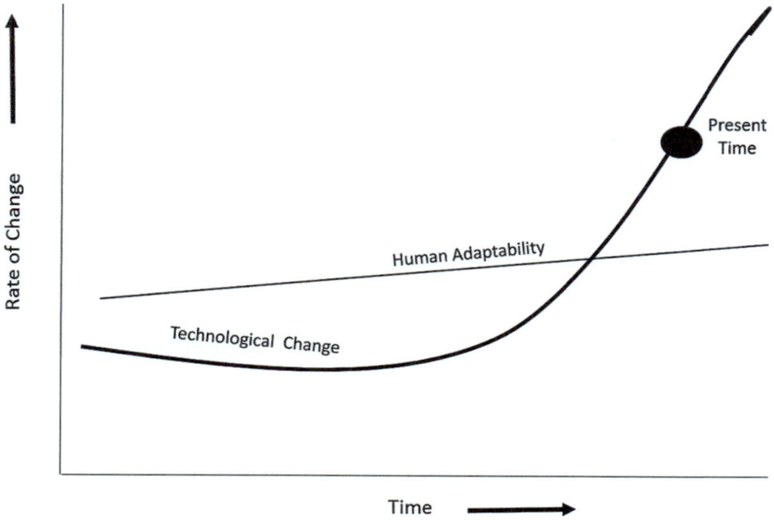

Teller's Graph: Modified from Thomas Friedman *Thank you for being late*, p.32

What of GIScience in terms of this hyper-paced, Fourth Industrial Revolution, where conflict and confusion give rise to division, loneliness, isolation and online tribalism? Obviously, the perspective presented here is one revolving around the domination of digital technology on our social fabric. Yes, GIScience has played and will continue to play a role in the mass migration towards a life dominated by all things digital – we are weaving our threads into that fabric just like everyone else, knowingly or not, for good or for ill. Is this the context that will control our directions in education and research? Will we succumb to the perfect storm? Have we spoken as a community about the context we wish to see, or how we can react to the context provided?

While the 1957 timeline had a distinctive and clear reason for being, and the 1995–2015 section focused on the intensive creation and pervasive consumption of digital technology (including data), this next transition is problematic for several reasons. This is a time of rising credentialism. We must be alert to how often we post new positions requiring a master's or Ph.D. qualification simply to gain entry to the marketplace in cases where they should never be required. It ends by diminishing the value of all qualifications (Collins 1979). It reinforces a time of 'us vs. them'. It has become a time of 'me'. There is no feeling that we are building something new, merely innovating something ephemeral. We have few dreams to latch onto and fewer goals to stand behind. When we are confronted with that reality, we retreat to our communities online in order to reinforce our anger and opposition.

We are more confused and more in conflict because we are allowing our community – our sense of community – to decline. This is as true for academic disciplines, such as GIScience, as it is for general life. We shift our identity and community as easily as we change clothes. The outcome of this is that no real community can be held together, because none of us has a desire to remain in any one, for any length of time. When there is no shared goal that can pull together such diverse and numerous communities, then we establish the perfect situation for more and continued harmful conflicts.

There are two reasons this is critical to grasp in terms of the current timeline and the overall goal of this chapter, which is to explore the future of what we do. The first is that we must focus on our goal (the only one goal), our 1957 moment. We are going to have to learn how to do more 'give and take' across our boundaries. In the bluntest of terms, dealing with the impacts and adaptations required due to climate change will be nothing less than the greatest undertaking, one focusing all of the collective energy and creativity of our species. Anything from the past, even from the most impressive undertakings like the moon shot, is small in comparison. We need to change, and in ways we do not even know, and over an immense period of time that no one has any experience with. It is not hyperbole. It is merely an extension of science from today forward over the next century. This is the reason for the choice of 2015 as the anchor for this section. The Paris Agreement from the COP21 process makes 2015 hold a special place. It is a reminder, perhaps even the start of a clock counting down.

The second reason is that 2015 (ish) was the moment when our use of social media, and the abuses of the social media empires, converged to create the perfect storm to which Neil Postman warned us in his classic *Amusing Ourselves to Death (Postman* 1984). He was not the first, and he recognized the influence of his colleague Marshal McLuhan who was certainly one of the first in analysing the impact of media on society. In more recent times, we have had Maggie Jackson (Jackson 2009), Sherry Turkle (Turkle 2011), Tim Wu (Wu 2016), Adam Alter (Alter 2018), Amy Chua (Chua 2018), Jean Twenge (Twenge 2017) and Carl Honore (Honore 2004), among numerous others. Just like the first scientists to warn us of climate change, these voices pointed out the logical end of our movement towards something we cannot grasp. To take a twist on McLuhan's idea that 'we march into the future looking backwards', this period is dominated by a disregard for the past. We are facing both a 'cognitive crisis' and a 'resistance to knowledge' at the very moment we need both the most (Gazzaley 2018).

Just when we need to focus, we are distracted. Just when we need to know the facts, we are presented with anything but factual information. Just when we need to talk across our social groups to find common ground, we continue to either ignore each other, yell over one another or spend more time reinforcing our ideas amongst other like-minded souls. Just when we need technologies that can allow us to do good, we are drowning in tools that are both addictive and intellectually, neurologically and physically destructive (Spitzer). When we need to have faith in a system that not only provided freedom but also educated for informed choices, we see Brexit and Trump and a host of other things we regret. Is it any wonder our faith and engagement in democracy are beginning to fall (PEW 2017)? Of course not; after all this is the age, unlike 1957, when even science is questioned.

These are not predictions. They are current happenings or things just about to reach our consciousness. The real surprises to come are not even in the lab yet. Our students have not thought up the great game changer to come in 20 years. Even so, this next period of time will see one of two things in terms of technology: a further leap into deeper integration of technical and human lives or a concerted effort to critically define what we want and need, instead of what is fun. If the average Internet user spends 6 h in front a screen, online, then we shall surely see more "digital dementia" arise (Spitzer 2012). It may come to pass that we cannot even recognize that tension and confusion between our lives and the screen (Forster 1909). Digital dementia will have become so pervasive that no one will even know what the term good-ole-days means, much less when they were. In the end, the fear from the Cold War will not come to pass; we are not going to blow each other up. We are more likely to simply ignore each other to the point that it won't make any sense to try.

There is no doubt that the future will include geospatial technologies and the data we need. It will not look the same; it will not be driven by geography. It will be driven by computing and data infrastructures where search engines will have to become more powerful because of the size of data. The only way around this is to halt or hold back some aspect of what we do. Is it possible to deal with limiting the number of new satellite platforms? What about the geosurveillance? What about predictive policing? What about health data? What about everything else? We are so needed to find solutions requiring more data. It is not just sad; it is ironic: we are seeing the culmination of our 1957 collaboration and our 1995–2015 creative period being undone by the very things we gave birth to. We often speak of pendulum swinging. This shift, and the speed of it, is more like the pendulum dropped.

The future is always good, bad and indifferent (apologies to Carl Sagan). The abilities we have, and the technologies, models, etc. make us not just the most valuable but the most vital. Everything to do with where and what's happening there, and how do we understand it, talks about it:

> We believe that if men have the talent to invent new machines that put men out of work, they have the talent to put those men back to work. John F. Kennedy

Conclusion

These final words were designed to be a reflection upon the state of that which surrounds our work. It is not and never will be a critique of any of the authors in this text.

There is nothing but admiration and envy for their work. This has been but another viewpoint, one from a person who loves the technology and uses it for the benefit of people through research and teaching – only a fool would suggest air and water are things never to be had. These are also the reflections of my very suspicious and sceptical nature. The more someone pushes something as the greatest and most essential thing I must have or buy, the more I must question it.

If anyone thinks we are in a normal time, then I ask that they seek out a few of the people I have as references at the end. Every time in history, technology, business and life is unique. However, this time is a period that confounds everyone. This is an age of populism, but it is not populism about leaders alone. We are in an age of technological populism. The problem with it, like political populism, is that is lacks a goal which is inclusive and which creates something better than ourselves rather than shrinking into our fears or weaknesses. It lacks a goal like the one established in 1957 or 1962. At the end of this time, the worst thing we can say about how we progressed is:

We have not done well. We've done lucky, and done it well.

Not everyone agreed with the space race either, but we accomplished it. Actually, we didn't. The generation born in 1930 did. That is a message. Those who came before us should be recognized, and we should pay attention to them. We are moving ahead without recognizing or appreciating the past. Such actions are the folly of an ignorant person who feels their views are new and unique, never seen or heard before. They need to find out their ideas, successes and developments are part of our communities over time and steps of others which are extensions of what was learned.

As we strive to create a zettabyte world, we will find we are in need of gigabyte lives full of megabyte wisdom. There remains a positive side to the future, of the next steps. The development of solid pedagogy does not change in value, even if content is altered or new creations come along. Developing ways to teach technology is unique, but the craft of teaching still has a foundation that goes beyond developing skills, helping students navigate the latest software or programming language.

Good teaching remains and will remain good teaching. The issue is that students (and society) may become more demanding about the technical, which is temporary, rather than the thoughtful and creative, which lasts and can be applied to anything and is an adaptive approach. In considering this, we need to look at the three things – our students (and ourselves), our psychology (and mental health) and the patterns of technology and the explicit goals of various entities be they spatial networks, or media, or government (tech-utopia).

There was no way we could have predicted what would happen at the time when GIScience began. We would never be able to predict what would have happened when Roger Tomlinson created the first GIS. To this day, I am reminded of, and live by, the words of my mentor and hero – give back all you can. (Boychuk 2014). One of the most prolific and influential people in our field, Mike Goodchild, has given us much to think about and build upon for years to come. He has shaped more of our field than anyone. Even so, we need to (as I think he would suggest) be very careful not to allow our egos and successes get in the way of doing good, doing what is right and being true to the nature of geography – to know and explain our world.

So what shall we do? We must find our place and follow our goal. The choice isn't hidden in fog or intellectual; it is clear and scientific. Our energies must be focused on climate change and all the pieces of that or that come from that. There is no room for delay, and every discipline and field of inquiry and every job and role everyone has must be relatable to that issue.

We can no longer be as we were in other stages of our development. We can no longer be doing it 'for the sake of doing it'. We created 'stuff' simply because we could, and this is no longer a valid premise. We innovate, but we have become stuck. We are in a funk. We don't even wish to think about it anymore. We have more than enough to keep us busy, but distraction is growing. We need more to solve the greatest challenge ever – ever! – first and foremost our community and our focus on contributing to reaching the goal.

Even so, I am always buoyed by the simplest of facts. The tin can was developed in 1810, while the can opener was invented 1855. Sometimes what we create comes along, and we feel it is right and good. Later on we find out we missed a piece, and then the truth becomes clear, and creativity comes to the fore. For good or bad, the date 2007 will likely become as recognizable as the moon landing or, for us, GIS. The iPhone will go down as the greatest and worst thing in technology. We love the perceived and unprovable convenience of our digital devices, and yet we still feel tied to them – not with a beautiful silk bow, but with cold shackles. I am still searching for the academic who loves email. I still hope the iPhone is the tin can, waiting for a useful opener to come along.

Everything has a context. *Everything*. What we wear has to do with the weather as much as fashion. What we eat is about where we are, what is available and the mood we are in. Context shines a light on what we do. It tells us what we may choose. It is not deterministic, and yet it does constrain our possibilities. This is true for everything. For education, every aspect of teaching, learning, research and application is supported by, or hemmed in by, the context surrounding the search for knowledge. Most errors or failures in education, as in any domain, can be traced back to a lack of deep and honest study of the context within which change was taking place. GIScience is not an island, standing alone and free from anything other fields feel as normal pressures. We are just as much subject to these things as others.

One could have dove head first into the details of specific questions like curriculum, pedagogy, credentials, professional education, the connection of research with learning and vice versa. All these and more are ripe for discussion – and thankfully the other authors have taken great care to make successful work of these in this text. That said, I feel better about today, at this moment in education and any 'cognitive crises', because we are reviewing the context of the past, peeking into the future, peering at the lessons to be learned and proposing those things we can and must do within GIScience which will help our field do more, for more. We may well be the most essential community that is needed to make our planet and our lives sustainable and more enjoyable. We only have one chance to get it right.

> 'But I can see you!' she exclaimed. 'What more do you want?'
>
> 'I want to see you not through the Machine', said Kuno.
> 'I want to speak to you not through the wearisome Machine'.

'Oh, hush!' said his mother, vaguely shocked.
'You mustn't say anything against the Machine'.

'Why not?'

'One mustn't'.

E.M. Forster's The Machine Stops.
The Oxford and Cambridge Review, 1909

References

ADL (1995). Alexandria Digital Geospatial Library. http://legacy.alexandria.ucsb.edu/adl/about_adl.html. Accessed 25 Mar 2018.

Alter, A. (2018). *Irresistible: The rise of addictive technology and the business of keeping us hooked*. New York: Penguin.

Amazon Technologies Inc (2016). System and method for transporting personnel within an active workplace. US Patent 9,280,157 B2. 8 Mar 2016.

Boychuk, R. (2014). Special to the globe and mail: Obituary Roger Tomlinson The father of computerized cartography. https://www.theglobeandmail.com/technology/science/roger-tomlinson-the-father-of-computerized-cartography/article17185506/. Accessed 24 Sept 2018.

Cassell, L. (2018). Everyone can code – a coding and computational thinking. https://www.teachontario.ca/community/explore/coding-in-ontario-classrooms/blog/2018/02/02/everyone-can-code-a-coding-and-computational-thinking-plc. Accessed 1 Sept 2018.

Cave, A. (2017). What will we do when the world's data hits 163 Zettabytes in 2025? Forbes. https://www.forbes.com/sites/andrewcave/2017/04/13/what-will-we-do-when-the-worlds-data-hits-163-zettabytes-in-2025/#5d137e74349a. Accessed 13 Oct 2018.

Chua, A. (2018). *Political tribes*. New York: Penguin.

Colbert, S. (2017). Billy Joel interview "The Late Show" with Stephen Colbert. Published on Jan 10, 2017. Accessed from https://www.youtube.com/watch?v=BEXQaxUjesE. Accessed 14 Jan 2017.

Collins, R. (1979). *The credential society: An historical sociology of education and stratification*. New York: Academic Press.

Dore, L. (2018). Amazon say patent to put workers in robot cages was a 'bad idea'. https://inews.co.uk/news/technology/amazon-robot-cages-workers-patent-robotic. Accessed 30 Sept 2018.

Duckham, M., Goodchild, M. F., & Worboys, M. (2004). *Foundations of geographic information science* (p. 4). Boca Raton, Florida: CRC Press.

Forster, E. M. (1909). The Machine Stops. In The Oxford and Cambridge Review (November). London: Archibald Constable. Retrieved from http://archive.ncsa.illinois.edu/prajlich/forster.html.

Friedman, T. L. (2016). *Thank you for being late an optimist's guide to thriving in the age of accelerations*. New York: Farrar, Straus and Giroux.

Gazzaley, A. (2018). The cognition crisis – Anxiety Depression ADHD Dementia. The human brain is in trouble. Technology is a cause and a solution. Medium. https://medium.com/s/futurehuman/the-cognition-crisis-a1482e889fcb. Accessed 18 Sept 2018.

Goodchild, M. F. (2007). Volunteered Geographic Information. http://www.ncgia.ucsb.edu/projects/vgi/docs/position/Goodchild_VGI2007.pdf. Accessed 4 Aug 2018.

Goodchild, M. F. (2009). Geographic information systems and science: Today and tomorrow. *Annals of GIS, 15*(1), 3–9. https://doi.org/10.1080/19475680903250715.

Gore, A. (1998). Towards a digital earth. Speech given at California Science Center in Los Angeles on January 31, 1998. http://www.digitalearth-isde.org/society/56. Accessed 20 Sept 2018.

Honoré, C. (2004). *In praise of slow: Challenging the cult of speed*. San Francisco: Harper.

IEEE. (2018). Ethically aligned design: A vision for prioritizing human well-being with Autonomous and Intelligent Systems (EADv2). Piscataway, New Jersey: Institute of Electrical and Electronics Engineers (IEEE). Retrieved from https://standards.ieee.org/industry-connections/ec/auto-sys-form.html.

Infosys (2016). Amplifying human potential: Education and skills for the Fourth Industrial Revolution. Accessed from: http://www.experienceinfosys.com/humanpotential.

Jackson, M. (2009). *Distracted: The erosion of attention and the coming dark age*. New York: Prometheus Books.

Kohn, C. F. (1970). The 1960's: A decade of progress in geographical research and instruction. *Annals Assoc Amer Geog, 60*(2), 211–219.

McKinsey (2018). Discussion Paper. Noes from the Ai Frontier. Jacques Bughin, Jeongmin Seong, James Manyika, Michael Chui, and Raoul Joshi. Accessed from: https://www.mckinsey.com/featured-insights/artificial-intelligence/notes-from-the-ai-frontier-modeling-the-impact-of-ai-on-the-world-economy.

McLuhan, M. (1967). *The medium is the massage*. London: Penguin.

PBS (2016). Interview with Kean Kranz. http://www.pbs.org/wgbh/nova/tothemoon/kranz.html.

PEW (2017). Globally, broad support for representative and direct democracy: But many also endorse nondemocratic alternatives. Richard Wike, Katie Simmons, Bruce Stokes and Janell Fetterolf. http://www.pewglobal.org/2017/10/16/globally-broad-support-for-representative-and-direct-democracy/. Accessed 12 Jan 2018.

Postman, N. (1984). *Amusing ourselves to death*. York: Methune Publishing.

Spitzer, M. (2012). *Digitale Demenz*. Munich: Droemer Knaur.

Turkle, S. (2011). *Alone together: Why we expect more from technology and less from each other*. New York: Basic Books.

Twenge, J. (2017). *iGen why today's super-connected kids are growing up less rebellious, more tolerant, less happy and completely unprepared for adulthood and what that means for the rest of us*. New York: Simon and Schuster.

Unwin, D. (Ed.). (2011). *Teaching geographic information science and technology in higher education*. London: Wiley.

WEF (2018). The future of jobs report 2018. World Economic Forum. https://www.weforum.org/reports/the-future-of-jobs-report-2018. Accessed 20 Sept 2018.

Wu, T. (2016). *The attention merchants: The epic scramble to get inside our heads*. New York: Vintage Books.

Index

S. Balram, J. Boxall (eds.), *GIScience Teaching and Learning Perspectives*, Advances in Geographic Information Science,
https://doi.org/10.1007/978-3-030-06058-9

H

MIX
Papier aus verantwortungsvollen Quellen
Paper from responsible sources
FSC® C105338

If you have any concerns about our products,
you can contact us on
ProductSafety@springernature.com

In case Publisher is established outside the EU,
the EU authorized representative is:
Springer Nature Customer Service Center GmbH
Europaplatz 3, 69115 Heidelberg, Germany

Printed by Libri Plureos GmbH
in Hamburg, Germany